Energy, Cold Fusion, and Antigravity
================================

Energy, Cold Fusion, and Antigravity

By

Frank Znidarsic P.E.

ISBN 9781-48027-0237

Revised 04/2026

Copyright © 2012 Frank Znidarsic

No bulk reproduction without written permission.

Limited copying for personal use is permitted.

Energy, Cold Fusion, and Antigravity

Table of Contents

INRODUCTION	4
CHAPTER 1 Energy	6
COAL	8
OIL	15
NATURAL GAS	17
RENEWABLE ENERGY	22
NUCLEAR FISSION	27
FUKUSHIMA	31
NUCLEAR FUSION	33
CHAPTER 2 The Genesis of Energy	37
THE CREATION	39
THE GENESIS PROCESS	46
CHAPTER 3 Electromagnetic Induction	49
GRAVITATIONAL INDUCTION	55
EMPIRICAL COLD FUSION	61
CHAPTER 4 Spacecraft Propulsion	68
CHAPTER 5 The Quantum Mystery	76
OUT OF FEYNMAN'S DARK ALLEY	80
Chapter 6 The Mathematics	84
THE SPEED OF HYDROGEN'S ELECTRONS	89
THE SPEED OF SOUND IN THE NUCLEUS	91
THE RADAII OF THE HYDROGEN ATOM	95
THE RADAII OF THE HEAVY ELEMENTS	99

Energy, Cold Fusion, and Antigravity

QUANTUM SPIN	101
THE SPECTRAL INTENSITY	106
THEORETICAL COLD FUSION	112
CHAPTER 7 Mysteries Resolved	114
THE PHOTOELECTRIC EFFECT	118
THE FINE STRUCTURE CONSTANT	122
THE DEBROGLIE WAVE	123
RELATIVITY	124
Chapter 8 Generations	140
Chapter 9 Conclusion	141
ABOUT THE AUTHOR	143
NOMENCLATURE	145
THEOREMS	148
PARTIAL BIBLOGRAPHY	149

Energy, Cold Fusion, and Antigravity

INTRODUCTION

Thank you for purchasing "Energy, Cold Fusion, and Antigravity". There are two schools of thought as to where we are heading. The first school of thought, as foreseen by climate scientist James Lovelock, predicts that we are rapidly heading towards a global catastrophe. The disaster will be triggered by the triple whammy of climate change, population increase, and the depletion of the fossil fuels. Lovelock's Gaia Hypothesis predicts that it is already too late. According to Lovelock, within this century, the only places on Earth that will be inhabitable will be at the poles. He advises, *"Enjoy life now while you can!"* Taking history as a guide, about 55 million years ago the Artic was a subtropical paradise and the tropics were too hot for most life. This period, the Paleocene-Eocene Thermal Maximum, arose after massive quantities of greenhouse gasses spewed out of volcanoes. The greenhouse gases emitted by the use of fossil fuels may be rapidly turning the earth's climate, once again, into a hothouse. The great intensity of some recent hurricanes may be a harbinger of the things that are to come. The aggregate of the effects have now pushed the earth into its sixth epoch of mass extinction.

A second school of thought, as foreseen by Peter Diamandis, predicts that we are heading into an age of abundance. This age will be ushered in by a rapid increase in the capabilities of computer and medical technologies. Fred Pearce in his book, "The Coming Population Crash", has revealed that societies that respect a woman's rights and provide for child care tend to maintain stable populations. Indeed the dire

Energy, Cold Fusion, and Antigravity

predictions presented by Paul Ehrlich, in his 1968 book, "The Population Bomb", did not come to pass.

Frank Znidarsic believes that the age of abundance can be acquired through the development of new, non-polluting, and abundant energy sources. The harnessing of the electromagnetic force has propelled us into this current age of wonder. New technology could classically harness all of the natural forces. These technologies will produce an abundance of clean energy. New propulsion technologies will efficiently propel spacecraft. These craft will provide resources from beyond the bounds of Earth. These technologies will propel us to the stars and lead us to the abundance of Peter Diamandis' singularity. "Energy, Cold Fusion, and Antigravity" places these technologies in a historical context. It goes on to describe the science behind a technology that will control all of the natural forces. These developments will expedite the coming of the next technological revolution.

Energy, Cold Fusion, and Antigravity

CHAPTER 1 Energy

The discovery of fire by ancient man, over 20,000 years ago, was a great achievement. Fire cooked his food and killed the parasites that afflicted him. He used fire, along with his new friend, the dog, to ward off predators. The carbon dioxide emitted by these fires may have resulted in a warmer and more habitual period. Bronze was first used to contain pressurized steam. In the 1st century AD, Hero of Alexandria invented the aeoliphile; a steam powered pinwheel. The aeoliphile demonstrated the potential of steam power; however, did not produce enough power to do any useful work. Higher pressure boilers came with the development of iron. In the 18th Century James Watt invented his steam engine. In 1803, Robert Fulton demonstrated the first piston powered steam ship. Charles Parsons invented the modern steam turbine in 1884. Parson's turbine converted large quantities of high pressure steam into mechanical power. The British navy was presenting before Queen Victoria's in the 1897 Diamond Jubilee. Parson's ship, the Turbinia, blew past the Royal Navy at a speed of over 34 knots. The Turbinia easily evaded a picket ship that attempted to catch it.

In 1824, Nicolas Léonard Sadi Carnot published the first theoretical understanding of steam power in his work; "Reflections on the Motive Power of Fire". Maxwell Boltzmann, extended Carnot's work, and linked the power of steam to the motion of molecules. Boltzmann's analysis presented some of the first theoretical evidence for the existence of molecules. Unfortunately Boltzmann was ahead

Energy, Cold Fusion, and Antigravity

of his time and the established scientists did not accept his ideas. Boltzmann committed suicide out of sickness and frustration. Steam power, according to Carnot and Boltzmann, was much like water power. The water pressure between the inlet and outlet of a water turbine corresponds to the temperature difference between the source and sink of a thermal engine. The temperature difference sets a theoretical bound to the efficiency of all heat engines.

$$Efficiency = \frac{Temp\ in - Temp\ out}{Temp\ in}$$

The pressures and the temperatures of steam boilers were increased as steel replaced iron and as welding replaced rivets. The increase in efficiency reached a climax in the 1960's. Temperatures of over 1000 degrees Fahrenheit were reached. At this temperature boilers convert 9500 BTU's of heat into one kilowatt-hour of electricity. The thermal efficiency of these units is about 35%. Pressures of three thousand pounds per square inch are needed to confine the hot steam. At this pressure the weight of steam equals the weight of water. Water no longer boils. The non-boiling liquid carries dissolved solids and minerals onto the turbine. The life expectancy of alloy steel diminishes at these high temperatures. The temperature, of the latest generation of boilers, has been cut back a bit in order to reduce material costs and increase reliability. Steam power still generates the bulk of our electrical energy.

Energy, Cold Fusion, and Antigravity

COAL

The combustion of wood and coal supplied the energy of the steam power. Coal still accounts for the bulk (about 50%) of U.S., electrical generation. My grandfather came through Ellis Island in 1916. His entry document stated that he was a coal miner. He came from a farm and knew nothing about mining. Coal was needed to power the industrial expansion of the time. I was told stories. The men were paid a commission on the coal they loaded. If the boss did not like you; he would give you a job cleaning up rock. There was no pay for the loading of rock. My father followed, in my grandfather's footsteps, and advanced to the position of a mine fire boss. In his time, machinery was beginning to be used in the mines. The operation of large, high powered machines, within the tight quarters of low Pennsylvania coal, was dangerous. The high powered machinery stirred up quite a bit of dust. My father suffered with black lung when he retired. I was the first in the family to attend college. I graduated with degree in Electrical Engineering in 1975.

I was told to get a job with a big mining company. Perhaps I would work mostly on the surface. This I did, and the pay was good. I was in the mine for a first time in Windber PA in the late 1970's. The mine was low and the miners crawled to a place to have lunch. I am tall, the roof was low, and I could not sit up straight. I tried to drink coffee out of my thermos. The coffee ran down the side of my face. I spied a high spot about 10 feet away. I thought, *"Those lazy, non-educated coal miners will not even bother to move to a better location!"* I took it upon myself and moved to the decent

Energy, Cold Fusion, and Antigravity

high place for lunch. That's when one of them asked, *"Found a nice high spot son! Do you know why it is high?"* I responded, *"It must have been dug out for some reason or another."* He replied, *"That's where the rocks keep falling!"* I had to crawl back under the low spot with the rest of the miners. I quickly learned to respect the knowledge of the experienced workers. Mining was hard, dirty work. I knew then that it was not stuck, by my heritage, into in my blood.

While in the mines, I noticed a lot of water dripping down from above. *"Where was it coming from?"* I asked, *"Probably from some farmers well or a small stream."* was the response. This was good clean water until it flowed into the mine and soaked up sulfur and other minerals. What happened to the crabs and lizards in the small stream? The mine's effluent would have to be treated for the next 10,000 years. Who was going to do that? In the future, they will surely hate us for leaving this mess!

Water collects in a basin, at the bottom of the mine, 300 feet underground. The first time I saw one of these basins, black water pouring in from all directions, I thought that I had discovered Dante's Inferno. The water was pumped up, by huge laboring pumps, from the basin to the surface. If these pumps could be run in reverse they would be mini hydro stations. My work, as an engineer, took me from the underground pumps up to the surface and into the water treatment plants. The process of cleaning up the bulk of the water was simple. The first step was oxidation. Air was blown through the water to oxidize it. The reaction converted the soluble green ferrous iron into a non-soluble red ferric iron. The pH was raised and the influent was

Energy, Cold Fusion, and Antigravity

seeded with alum. The ferric iron precipitated out (due to the common ion effect at the high pH) with the alum in the clarifier. The effluent of the clarifier still contained traces of many elements including mercury, boron, and selenium. The removal of these trace elements dramatically increased the cost of the water treating operation. Anaerobic bacterial digesters are required to remove selenium. The bacteria live in, what was called, the bug tank. Tighter requirements will require the use of activated carbon final filters. Who is going to keep this equipment running for hundreds of years? The plugging of abandoned mines, with lime slurry, was considered. It sounds like a good idea; however, it may not be possible to plug all of the fine cracks.

Modern strip mining replaces the coal's overburden and the top soil. Strip mining does not contaminate the streams to the extent that deep mining does. Acidic rocks lie just beneath the surface of the reclaimed land. These rocks tend to kill deep rooted trees. Locust trees, Japanese knotweed, and elder take up residence on the reclaimed land. The land does not appear to be of much use in farming. Coal mining leaves a legacy of bad water and bad land. Mining operations create a lot of jobs. These jobs create wealth at the expense of the environment.

I was hired, as an engineer in a large coal fired, power plant in the 1980s. I felt bad about leaving Chief Mining Engineer Ray Frank. He had spent four long years painstakingly teaching me so much. I was now working full time on the surface, however, I still dream about Ray telling me to do something or another. In the 1980's, power companies would often showcase their smoke stacks. *"Take a look!"*

Energy, Cold Fusion, and Antigravity

they would say, *"You can hardly see any soot coming out of the stack!"* They had a shaded opacity reference chart. The stack's low opacity, near the bottom of the chart, was a testament to the wonders of modern clean coal technology. I asked. *"Why are the tree hugging extremists making such a fuss about the little bit of brown smoke that remains?"* One of my assignments was to automate an electrostatic participator's pneumatic ash removal system. I did a fine job of it. It would automatically enter into a default mode if any single piece of equipment failed. The power plant would keep running at a reduced load. I was a true believer in the clean stack hypothesis.

Others, more informed than I, were not impressed by the singular performance of the electrostatic participator. It did not capture that little bit of brown smoke. Contained in that smoke were the noxious gasses. Nitrous oxide was produced in the boiler as the nitrogen in the combustion air reacted with oxygen. The reaction proceeds within the high temperature of the boiler's flame. Smog formed as sunlight reacted with the released nitrous oxide. New regulations followed that required the reduction of these nitrous oxide emissions. In this effort, the temperature of the boiler's flame was reduced. The flame temperature was reduced by restricting the combustion air that was supplied to the flame. Boilers, up to that time, were designed to use an exact (stoichiometric) fuel air mixture. The coal did not completely burn under the low oxygen, low NOx, condition. Some of pulverized coal carried over to the fly ash. The loss of ignition (LOI) reduced the boiler's efficiency and complicated the disposal of the, now flammable, fly ash. It

Energy, Cold Fusion, and Antigravity

was a balancing act to keep these low-NOx burners operating correctly. These burners, alone, could not meet the increasingly tighter regulations. Selective Catalytic Reducers (SCRs) were now required. They reduced the nitrous oxide in much the same fashion as does an automotive catalytic converter. These converters run hot, like the one in an automobile. Sections of boiler's economizer heat exchanger tubes were bypassed to keep the temperature of the exit gas up. This reduced the boiler's efficiency by 15% or more. A plant superintendent summed it up best with his comment, "*A perfectly good boiler was ruined!*"

The coal comes with a few percent of sulfur (S). Sulfur reacts with oxygen, in the boiler, and forms sulfur dioxide. The sulfur dioxide (SO_2) reacts with the water vapor in the atmosphere and comes down as sulfuric acid rain (H_2SO_4). New laws required sulfur dioxide to be removed from the flue gas. A pretreatment of the coal was tried. The coal was crushed, microwaved and washed. This low cost technique was not effective. Sulfur dioxide had to be scrubbed out, after combustion, in a limestone solution. These limestone scrubbing systems are mammoth and expensive. They cost as much as 30% of the original plant and consume about 4% of the plant's electrical power. They have proven to be, in spite of these costs, economic. The company I worked in the 21st Century built excellent scrubbers of all types.

Carbon Dioxide (CO_2) is the main product of combustion. Like a blanket, it reflects thermal radiation back down to the Earth. The reflected energy heats the Earth. Carbon dioxide is the next gas to be removed from the exhaust stream. Plans are in the works to tax carbon dioxide emissions. Carbon

Energy, Cold Fusion, and Antigravity

dioxide constitutes 15% of the flue gas. That's a lot of gas. Its removal makes for a big material handling problem. Carbon dioxide dissolves in a liquid in the same way that it's does in a bottle of soda. It goes into solution in cool water and produces carbonic acid. The same process drives carbon dioxide into the sea. Excess carbon dioxide may turn the sea acidic and septic. Hydrogen sulfide gas is emitted from a septic condition. It's the gas that gives a sewer its smell. Hydrogen sulfide is poisonous. Some say that the hydrogen sulfide that was emitted, by a septic sea, killed the dinosaurs. God help us if that happens again! The flue gas is cooled to drive the carbon dioxide into solution. The carbon dioxide is then driven out of the solvent by the application of heat. The heating of the sea, by a global warming, will also drive heat trapping carbon dioxide out of solution. The increase in atmospheric carbon dioxide produces more heating. The result is a runaway increase in global warming. The cycling, of the temperature of the solvent, is an energy intensive process. The process of capturing carbon can consume 25% of the plant's electrical output. In conjunction with the other emission controls over 30% of the plant's output can be consumed. For every three plants that are equipped with every clean technology another has to be built to supply the parasitic load of the environmental technology. More coal must be mined to supply the parasitic load. Clean coal technology has reached the point of diminishing returns on environment and capital. Newer and more efficient solid carbon dioxide sorbents are now being tested; however, no magic bullet is yet in sight. The capital costs of a carbon sequestering facility are huge. They can equal the cost of the original plant. The carbon dioxide must be sequestered

Energy, Cold Fusion, and Antigravity

underground at high pressure (5,000 PSI). A gas bubble, 25 miles in diameter, can form. Who is going to insure the unknown liability associated with this giant gas bubble? What will the mineral rights for the disposal cost? If power plants are lucky enough to lie near an oil field, the carbon dioxide can be pumped into an existing oil well. Carbon dioxide acts and solvent and helps to extract the residual oil. The use of the carbon dioxide, in this application, makes good sense.

Blow down from the scrubber, leachate from ash piles, and runoff from the coal piles require additional water treating facilities. Senior power plant engineer Dave Jacoby stated, *"All of this, just to spin the generator's shaft!"* Dave never saw the heroic efforts in the mines. Utilities are now backing away from investment in new coal fired power plants. It is not practical, in many cases, to meet the, December 2011, Mercury and Air Toxics Standers act. Many older coal fired power plants are now being shuttered. The United Mine Workers of America has estimated that 56,000 megawatts of coal fired capacity will be retired. In November of 2012 the Union of Concerned Scientists released its study. They found that as 353 generators in 31 states, totaling 59 gigawatts of generation capacity will likely be too expensive to operate after the installation of the additional pollution control equipment. These potential closures are in addition to 41 GW of already announced coal retirements. Coal was to have reigned supreme for the next 200 years. King coal appears to be headed for a hard time. The dream of an energy independent, coal fired economy is gone.

Energy, Cold Fusion, and Antigravity

OIL

Oil is the most precious of the fossil fuels. It is packed with energy. It is a liquid at room temperature and it fuels the world's transportation system. Oil powered trucks transport coal to the power plants. Roads are built with the asphalt that was extracted from oil. Renewable energy sources, such as wind and solar, require the plastics that were made from oil. Crops are fertilized, harvested, and transported with oil powered equipment. In the 1960's, vast quantities of oil spewed out of the wells in Saudi Arabia. The stratum was saturated with over 70 percent by weight of oil. The sweet crude was priced at one dollar per barrel. An energy intensive, good life was to be had. The thing to do, in the 1960s, was to get a bigger engine and burn rubber!

M. King Hubbard presented a paper to the American Petroleum Institute in 1956. The paper correctly predicted that oil production, in the United States, would peak in the late-1960s. Hubbard went on to predict that global oil production would peek after the turn of the 20st Century. This peek has been delayed a bit by better extraction techniques. The world's oil supply is now peaking. The world's oil supply will soon enter into a slow state of steady decline. The world's greatest oil field, the Ghawar Field in Saudi Arabia, was placed on line in 1950's. Pressurized gas blew the oil out of subsurface cracks and up to the surface in a gusher. As these fields became depleted as the gas pressure was released. The released gas collected in a pocket above the oil. Many of the oil fields in Saudi Arabia now have useless gas caps. The oil no longer blows out of the crevices

Energy, Cold Fusion, and Antigravity

in the ground. New techniques, such as horizontal drilling, fracking with high pressure water, fracking with small explosive charges, and the flushing of the crude with water can extract oil from strata containing less than one percent by weight of oil. These techniques, if pushed too hard, can leave unrecoverable pockets of oil in the ground. The techniques tend to quickly deplete the remaining oil. Similar techniques have temporarily increased production in the U.S. A surprisingly large amount of oil is now being extracted in North Dakota and Texas. Unfortunately, most of earth's oil is located in countries that oppose western hegemony.

Oil is extracted from the tar sands in the United States and Canada for a price of about $50 barrel. The extraction process strips the oil out of the sand with heat. The heat source adds carbon dioxide to the atmosphere. The mining operations disturb massive amounts of ground. Many environmentalists are opposed to the extraction of oil from tar sands.

Gasoline engines require 20,000 BTUs of thermal energy to produce one kilowatt-hour of power. Their thermal efficiency is a dismal 17%. Government mandates have required more efficient gasoline powered engines. This increase in efficiency comes with an increased cost. The head of the Environmental Protection Agency, Lisa Jackson, resigned in 2012 as economic interests pushed back this environmental agenda.

Despite some temporary gains in efficiency and production, the end is coming. The earth is slowly running out of oil.

Energy, Cold Fusion, and Antigravity

NATURAL GAS

In the late 19th Century, coal gas was used in lamps. Carbon monoxide was produced by restricting the oxygen supply to burning coal. Carbon monoxide has an energy content of about 125 BTUs per cubic foot. Early town gas contained a lot of carbon monoxide. Carbon monoxide is extremely poisonous. Some old movies used the poisonous town gas in a homicidal plot. The hero saved the victim, a beautiful unsuspecting young lady, and became the love of her life.

The energy content, of coal gas, can be increased to about 300 BTUs per cubic foot by removing the nitrogen gas from the combustion air. Nitrogen constitutes 80% of the atmosphere. Nitrogen is practically inert and, as such, it does not react or dissolve. It could only be removed by, the energy intensive process, cryogenic distillation. The removal of nitrogen from the combustion air, to increase its BTU content, is an expensive proposition. The injection of steam into smoldering coal produces a chemical 'shift'. This reaction converts carbon monoxide into hydrogen and carbon dioxide. The carbon dioxide is then scrubbed out of the flue gas with a caustic solution. The remaining blue gas (blue burning without incandescent soot) has an energy content of about 600 BTUs per cubic foot. Methane (**CH_4**) has an energy content of over 1,100 BTUs per cubic foot. Natural gas is mostly methane and it is a good fuel. The discovery of new supplies of conventional natural gas peaked about 40 years ago. These conventional supplies are, as per Hubbard's has predicted; now depleting. In the early 1980s natural gas was in short supply. I worked in the steel industry in

Energy, Cold Fusion, and Antigravity

Birmingham Alabama. Processes were being studied that could produce 600 BTU gas from coal. The gas would be used to supply the existing heat treating ovens. An experimental plant produced gas, with an energy content of 1,100 BTUs per cubic foot, in Indiana PA. The faculty resolved many problems and proved the technology. The gasification plant could produce a high quality gas at a competitive price. The price of natural gas dropped in 1990 and the experimental gasification facility was shut down, and scraped.

Gasification technologies are now being resurrected. Coal gas would be pretreated before combustion, in a power plant. The pretreatment would eliminate the expensive post combustion sulfur, mercury, and carbon separation systems. A marketable byproduct of the cryogenic distillation of air is argon gas. The costs of coal gasification, pre-treatment, and combustion are not proven. I expect that the costs associated with this technology, such as the distillation of the nitrogen out of the combustion air, will be high.

New supplies of natural gas have been, unexpectedly discovered in Marcellus Shale. The development of these supplies began in 2004. In 2011, 16,000 new gas wells were drilled in the U.S. This gas lies one mile beneath the surface. The drilling turns horizontality at a depth of one mile. It then proceeds horizontally for another mile. The shale is fracked by driving millions of gallons of sand and chemicals (including the carcinogen benzene) into the wells. The fracking process cracks the shale and liberates the natural gas. The fracking water soaks up the usual suspects; sulfur, organic material, heavy metals, and salt. Once used, the

Energy, Cold Fusion, and Antigravity

fracking water has been transported to municipal sewage treatment facilities for disposal. These faculties employ bacteria to digest and remove organic material. These facilities can remove benzene in concentrations of up to 50 parts per million. They do nothing to remove the dissolved solids. These solids may, in fact, kill the bacteria and upset the sewage treatment plant. The only way to clean up the diverse waste streams, associated with fracking, is through a process of distillation or a process of reverse osmosis. Distillation is energy intensive. The membranes, used in reverse osmosis, tend to foul. More recently the franking water has been partially cleaned and reused in other wells. This solution has greatly reduced the amount of discharge.

The contamination of local supplies of ground water has been reported. Small earth quakes may have been initiated, by fracking, in the Midwest and in Ohio. The deep natural gas contains high quantities of radium. France has banned fracking. The United States has started cracking down upon the industry. New legislation intends to regulate the gas that escapes from the bore hole when the bit is removed, require a full disclosure of the fracking ingredients, and issue tough water discharge mandates.

Liquid carbon dioxide is now being introduced as a fracking fluid. The cryogenic fluid is quickly injected into the well. It vaporizes as it contacts the subsurface strata. The gas greatly expands and efficiently fracks the shale. The carbon dioxide does not remain in the ground, it cannot contaminate the ground water, and it does not adsorb minerals. The injected carbon dioxide emerges with the natural gas at the well head and it must be removed before the natural gas can be put into

Energy, Cold Fusion, and Antigravity

a distribution pipeline. Carbon dioxide has conventionally been removed by dissolving it in a caustic solution. A molecular sieve has recently been constructed from a new type of plastic. The sieve efficiently absorbs carbon dioxide from the well effluent without the use of hazardous chemicals. There are, of course, the additional costs associated with the carbon dioxide, maintaining the gas pressure so that the fractures do not close, the separation facility, and the transport of the gas mixture to the separation facility. It's going to interesting to watch the development of these situations over the next few years.

The benefit of the natural gas greatly exceeds its liabilities. Natural gas contains hydrogen. Hydrogen produces water vapor, not carbon dioxide, when burned. Water vapor forms clouds that decrease global warming. The latest generation of internally fired combustion gas turbines, with heat recovery, boilers consume only 7000 BTUs of thermal energy per kilowatt-hour of electrical energy. The thermal efficiency, of these units, is as high as 48%. Factoring in the parasitic loads associated with the latest generation of coal fired plants; a gas fired unit delivers twice the electrical energy per unit of thermal power. The increased efficiency reduces the emission of carbon dioxide by one half.

"Heat rate means a lot!" Jim Conway, Senior Engineer.

As an additional benefit Ethylene can be extracted from the natural gas. Ethylene is used to make plastics. One of these plastics is acrylic. Acrylic is extremely durable. Acrylic is now being introduced as a paint additive. This paint retains its finish. The storage facilities in the North East are

Energy, Cold Fusion, and Antigravity

currently filling with this new gas. At the time of this writing, the price of gas has dropped from $12 to $3 per 1,000 cubic feet. Given a BTU content of 1,000 BTUs per cubic feet, gas power costs about $3 per million BTUs. On a BTU basis (12,500 BTUs per lb.) the cost of natural gas equals that of coal at $75 per ton. The cost of deep natural gas has approached that of coal on a BTU basis. Plans to export the gas will drive up its price. The price of natural gas is expected to stabilize at $5 per 1000 cubic feet. The cost of gas is expected to remain low enough to stymie investment in new coal fired generation. Utilities are turning away from coal and to natural gas. The estimate of the amount of extractable natural gas has been updated by a factor of three. Some geologists believe that even larger supplies of gas were trapped deeper within Earth during its formation. The primordial gas is of an inorganic origin. To date, none of this primordial gas has been discovered. A limitation of natural gas is that it does not liquefy at ambient temperatures. It must be compressed when used for transport. Compressed gas is much less dense than a liquid fuel. The low energy density, of a compressed gas, limits its use as a transportation fuel.

As the earth's natural resources are depleted we have become better at extracting the remains. The president of Black Sheep Technology, Ronald Anderson, has stated that the window of opportunity, provided for by new natural gas, will close in 50 years. If new limitless sources of energy are not on line by then, there will not be enough energy left to facilitate a transition. If Ron is correct, it may be back to the Stone Age for all of us.

Energy, Cold Fusion, and Antigravity

RENEWABLE ENERGY

Water power requires vast amounts of water and a high drop. The best sites, in the U.S., were developed decades ago. The building of the great dams, in the Western United States, was followed by a rapid population increase. The West has now become dependent upon the dammed up water. The onset of global warming may have produced a drought that limited the flow of water into these dams. This is a bad omen. In 2010 the giant Three Gorges Dam, in China, was filled. One and a half million people were displaced by the flooding. Valuable farm land was lost. The Yangtze River Dolphin appears to have been driven to extinction. More recently landslides, triggered by the new lake, have displaced an additional 100,000 people. The dam was built to supply 10% of China's electrical power. Due to the rapid growth of China's economy, the dam now supplies only 1% of its power. The dam did not, as intended, reduce China's carbon footprint. Hydroelectricity cannot supply enough energy to power world's rapidly developing economy.

Many windmills have been erected across the country. Wind mills only produce power when the wind velocity is greater than 9 MPH. A wind mill spinning in calm weather may be motoring and using power. At speeds of more than 29 MPH they shut down to protect the generator. The production, of these windmills, had been supported by a tax credit of 2.3 cents per kilowatt. In late 2012 this tax credit expired. Layoffs ensued at the wind mill construction companies. It appears that wind energy cannot stand on its own merit. Similar economics apply to active solar power.

Energy, Cold Fusion, and Antigravity

The production of grain alcohol (ethanol) as a fuel is marginally viable. The distillation process requires a significant fraction of the product's energy. Ethanol corrodes lines, eats gaskets, pits aluminum, and is hydroscopic. A few local farmers have tried to make ethanol for their own use. They discovered that the Ethanol destroyed their smaller engines and that it just was not worth the all of the trouble. Recent droughts have shown that food production is being effected by climate change. Ground water is being rapidly depleted in order to keep up the production. Ethanol production facilities are now being shuttered. There is hope that genetically engineered crops may enable bio-diesel (vegetable oil) to reach economic viability. The cultivated fields sequester much less carbon dioxide than natural grasslands. Is the diversion of farm land, natural land, and food to the production of fuel ethical?

Efforts are also underway to develop an efficient technology to convert cellulose (stalks, grass, and other waste) into ethanol. Existing technologies break down the cellulose with high pressures, high temperatures, and sulfuric acid. They are energy intensive and expensive. Termites and fungi slowly break down cellulose but they do not excrete ethanol. There is an ongoing effort to take genes from termites and fungi and to implant them into yeast. We hope that such a genetically modified bio-organism and its associated enzyme does not get loose and digest all of the rubber gaskets, trees, and the rest of our stuff as it did in the movie "The Andromeda Strain". Don't worry, if the fermentation of cellulose to ethanol was energetically practical nature would

Energy, Cold Fusion, and Antigravity

already be doing it. Taking this lead from nature; synthetic fuel technologies will not be easy or cheap.

Cost effective energy storage systems would make it possible to utilize the existing energy supply more efficiently. In the late 19th Century compressed air, flywheels, and mechanical springs were tired in an effort to motorize street cars. The range of these street cars was measured in a few thousand feet. The compressed air tanks exploded, the fly wheels blew apart, and the springs released at once driving a street car to destruction. Chemically stored energy is an order of magnitude more powerful than mechanically stored energy. Batteries store energy chemically. Since the turn of the 20th Century, attempts have been made to harness the possibilities of chemically stored energy. Thomas Edison introduced the nickel-iron battery at the turn of the 20th Century. It had a long life and could withstand many deep discharges. Unfortunately, it could not supply enough current to drive a traction motor. The nickel-iron battery tended to discharge itself over time. A modern nickel-iron battery stores about 40 watt hours of energy per kilogram of weight. The Exide Battery Company introduced the Exide lead acid battery in competition with Edison. It contained a little more energy than the Edison's battery on a per weight basis. It could deliver the high currents that are required by large traction motors. Unfortunately, it did not hold up for long after it was deeply discharged. Did you ever try to use a computer U.P.S. supply as a portable power source? Manufacturers incorporate a lockout to prevent this use. This type of use would deeply cycle the enclosed lead acid battery and ruin the battery. The lockout was added to prevent battery failure.

Energy, Cold Fusion, and Antigravity

Gasoline contains 13,000 watts hours of energy per kilogram of weight. It is no wonder that batteries fell out of favor after the introduction of the gasoline engine.

An Electrically Powered Fiero at Black Sheep Technology

It is having its electronic motor controller updated. The new controller contains Insulated Gate Bipolar Transistors. The transistors chop up the current and efficiently power the traction motor. The car has lead acid batteries in the trunk, under the hood, and in the back seat. With all of the batteries it weighs over 3,000 pounds. In spite of all of the batteries and the new controller, the car's range is limited to about 30 miles. The batteries are deep cycled and need to be replaced every year or so. General Motors recalled and scraped its lead acid electrically powered vehicle; the EV-1.

Lithium ion batteries are beginning to expand the possibilities of battery powered transport. These batteries deliver about 120 watts per kilogram of weight. That's still an order of magnitude, by weight, less than gasoline. The

Energy, Cold Fusion, and Antigravity

life of the lithium ion batteries is reduced by deep cycling; however, they can withstand deep cycling better than lead acid batteries. Batteries produce little waist heat. The heating and cooling load must be drawn electrically from the battery. Cold and hot temperatures reduce the battery capacity by as much as 60%. General Motors took a risk when they introduced the Chevy Volt. General Motors gambled that the battery technology would catch up.

I recently looked a motorizing my bicycle. A 50 cc 2 cycle gasoline conversion kit costs about $150. A 50 cc 4 cycle engine conversion kit costs about $350. The tanks on these retrofits store about a half a gallon of gasoline. These bicycles have a range of about 70 miles per full tank. A lithium ion battery conversion kit costs about $900. These bicycles have a range of about 15 miles. It takes time to charge the batteries and the expensive $500 batteries have to be replaced every few years. Given these economics, the production of the Chevy Volt has been suspended and A-123 Battery Systems has declared bankruptcy. Battery powered vehicles are getting better; however, they need to improve by a factor of ten to become competitive with gasoline powered vehicles.

"It really begs the question as to whether the electric car at its current level of development is a suitable means of transport for the modern age."

Dr. Barry Culpin, DeMontfort University

Renewable energy does not have the power to take us to the stars. It will not change the type of our civilization.

Energy, Cold Fusion, and Antigravity

NUCLEAR FISSION

The neutron was discovered by James Chadwick 1932. With this discovery man extended his knowledge of the particles that formed the nucleus. The impetus of moving neutrons was employed to split atomic nuclei. Ten years later, at the University of Chicago, the first nuclear chain reaction was produced. Developments followed rapidly. In 1954, Lewis Strauss, Chairman of the Atomic Energy Commission, forecast that nuclear energy would be, *"too cheap to meter."*

Major problems surfaced quickly. At 9:01 pm January 3, 1961, the U.S. army's Stationary Low Power Reactor #1 (SL-1), near Idaho Falls, exploded. The Atomic Energy Commission reported, "The #2 crew member was struck on his back and legs with water and/or steam causing him to be thrown against a shield block and landing in the vicinity of the instrument wells. The #1 crew member was also struck with water and/or steam and was thrown back against another shield block striking his head first. Simultaneously, the shield plug assembly impaled the #3 crew member and pinned him to the floor of the fan room, a distance of approximately 13 feet above the reactor's head." The plug assembly had entered the groin of operator Richard Legg, exited through his shoulder and propelled him straight up to the ceiling. There he remained impaled on the ceiling. It took six days to extract his body. The intense radioactivity killed the bacteria and preserved his body. What remained of the three crewmen was buried in lead lined caskets. An unknown amount of radiation was released into the surrounding farmland. The incident ended the U.S. Army's

Energy, Cold Fusion, and Antigravity

nuclear program. Several things went wrong. The supervisor, Richard Legg, engaged in horseplay. The wife of another operator, John Byrnes, just called and asked for a divorce. No one knows how these events may have contributed to the accident. Design flaws allowed the reactor to quickly become supercritical.

New reactor designs were introduced. Bubbles in the boiling cooling water would reduce the neutron moderating effect of the water. This reduction would tend to shut down the reactor. The removal of a single control rod would no longer release a critical reaction. Admiral Hyman G, Rickover tightened up the Navy's operator training and selection requirements. Rickover was tough; he did not allow casual conservation in the control room. Horse play, for sure, would no longer be tolerated. The lessons learned from the accident would be enforced. No more accidents were to follow. One of my biggest problems, on the job, has also been dealing with people. People have flaws, even at a nuclear plants.

On October 5, 1966 Fermi#1 nuclear plant suffered a partial meltdown. No radioactive material was released. The plant is located halfway between Detroit Michigan and Toledo Ohio. John G. Fuller in his book "We Almost Lost Detroit" stated that the accident came close to becoming a major nuclear disaster. John F. Fuller defined this disaster as a release of half of the reactor's core material. He wrote that, sooner or later, a nuclear reactor would suffer this fate. The release of the half of the core material could contaminate an area the size of Ohio for 1,000 years. Lessons were learned from the Fermi accident. It was a fast breeder reactor that

Energy, Cold Fusion, and Antigravity

employed liquid sodium as a coolant. Sodium violently explodes upon contact with water. The design was inherently dangerous. It took four years to repair the Fermi Nuclear Station. Its performance remained poor and, in 1972, the reactor's core was dismantled and decommissioned. Breeder reactors were subsequently abandoned for all but experimental purposes. America's effort at operating a full-scale breeder reactor had failed. The United States was now dependent upon finite supplies of fissile uranium 235. The light water reactors, that used uranium enriched with U235, were deemed to be much safer than a breeder reactor. It was believed that John F. Fuller's disaster, involving a release of half of the core material, was nearly impossible with a U.S. designed light water reactor!

A boiler feed pump tripped at the Three Mile Island nuclear plant at 4AM on March 28, 1979. The plant automatically shut down; however, a safety valve remained stuck open. The plant's operators misinterpreted the effect of the sticking safety valve and switched off the reactor's emergency cooling system. This action resulted in the melting of the reactor's core. A mass of hydrogen built up in the containment building. Some feared that the hydrogen would explode and John G. Fuller's disaster would occur. Pennsylvania could be uninhabitable for a Century. The containment building did not explode and the failsafe redundancy of the U.S. designed light water reactor was vindicated.

Foamed plastic, covered with two coats of a flame retardant paint was used as fire stop at the Browns Ferry Nuclear Power Plant. On March 22, 1979, one year after the Three

Energy, Cold Fusion, and Antigravity

Mile Island accident, a worker used a candle to search for air leaks. He accidentally set the material on fire. The fire spread from the foam to the reactor's control wiring. It caused a significant amount of damage. Once again the U.S. designed light water reactors were vulnerable in another way. The Nuclear Regulatory Commission reviewed the accident and implemented procedures that would prevent another such incident from reoccurring.

The Chernobyl nuclear plant, in the Ukraine, exploded and caught on fire. A large amount of the core material escaped. Entire towns became inhabitable. The Union of Concerned Scientists estimated that the number of additional cancer deaths from the released radiation will approach 25,000. In 2012, the president of Ukraine announced that an additional $1.58 billion will be spent on the sarcophagus that holds the remains of the dead reactor. Chernobyl's reactor employed non-enriched uranium fuel and flammable carbon as the neutron moderator. It had no containment structure. It was believed that such a disaster could not happen with better designed light water reactors.

Energy, Cold Fusion, and Antigravity

FUKUSHIMA

Fukushima is a U.S. designed light water reactor with containment. Unlike Three Mile Island, its containment structures did explode. An inspection of the reactor's core, with cameras, revealed that a significant amount of core material had escaped. Japan was very lucky that most of the radioactive material blew out to sea. John F. Fuller's disaster did occur just as he said it enviably would. The Fukushima disaster effectively bankrupted, TEPCO, one of the largest power companies in the world. It will cost up to a trillion dollars and take 40 years to clean it up the mess. It is an ecological disaster of unknown proportions. A significant fraction of Japan's land area, 4,000 square miles, or 0.3% of all the land in the country, must be abandoned for up to 50 years. 90,000 people are displaced and homeless. They have lost their farms, businesses, schools, livestock, and all other possessions. All, but two, of Japan's nuclear power reactors were shut down. Eighteen percent of the Japans' electrical generation was shuttered. The nuclear power industry in Japan was destroyed. Japan's economy is suffering. The U.S. designed light water reactor was much more vulnerable than most experts believed possible.

Japan's former Ambassador to Switzerland, Mr. Mitsuhei Murata, spoke at the Public Hearing of the Budgetary Committee of the House of Councilors March 22, 2012. Ambassador Murata stated that if the crippled Unit #4 reactor building collapses it will disrupt the common spent fuel pool. This pool contains 6,375 fuel rods. The radioactive rods are not enclosed by a containment structure. The consequence of

Energy, Cold Fusion, and Antigravity

the escape of this material into the environment would be horrific. The release of radioactive cesium, from the spent fuel pool, is exceptionally bad. Biological organisms tend to absorb and concentrate cesium and other radioactive elements. There is currently no long term solution to the disposal of radioactive waste. The Obama administration shut the planned Yucca Mountain disposal facility down. All nuclear plants must store large quantities of radioactive material on site. This has been shown to be a dangerous arrangement.

New passive reactor designs are now entering the market. These reactors can shut down safely with the loss of all electrical power. It has been stated that a release of core material is extremely improbable with this design. Many have lost their faith in this old promise. Will the next generation reactors be vulnerable; perhaps in another way? Professor Michio Kaku has suggested that nuclear reactors may be susceptible to the effects of a large solar storm. Germany and Japan are in the process of abandoning conventional nuclear power.

Energy, Cold Fusion, and Antigravity

NUCLEAR FUSION

In 1934, Ida Noddack introduced the concept of atomic nuclear fission. In December 1938, Otto Hahn and Fritz Strassmann bombarded elements with neutrons in their Berlin laboratory. They found elements that transmuted during the neutron bombardment. Uranium nuclei broke into two roughly equal pieces. The fracturing produced large quantities of energy. Within ten years of their discovery the first nuclear reactor Chicago Pile-1 was producing thermal energy. Fission involves the fracturing of the uranium 235. The earth contains a limited amount of uranium 235. Nuclear fission was first used to power atomic bombs. These bombs have a limit in their explosive yield of about 20 kilotons of TNT.

Nuclear fusion (the opposite of fission) bonds the light elements deuterium (heavy hydrogen) and tritium (doubly heavy hydrogen) into helium. The element deuterium is contained in sea water in a ratio of one part per 6,000. The supply of deuterium is virtually without limit. Nuclear fusion occurs when charged nuclear particles impact each other with enough impetus to overcome their mutual electrostatic repulsion. The rate, of the fusion reaction, is set by a combination of temperature, time, and pressure. This combination is called the Lawson Criteria. Nuclear fusion heats the sun. The surface of the sun is not hot enough to initiate a fusion reaction. Fusion occurs in the sun's core at a temperature of about 15 million degrees K and at a material density of around 160 g/cc. Under these extreme conditions, fusion progresses slowly over periods of billions of years. A

Energy, Cold Fusion, and Antigravity

man-made thermonuclear power plant must consume its fuel more quickly. These machines must run at temperatures ten times hotter than the interior of the sun. That's a daunting prospect. The first fusion devices were, Soviet built, layer cake atomic bombs. A very hot atomic fission explosion initiated a secondary fusion reaction. The resulting fusion reaction produced many neutrons. These neutrons initiated more fission. These fission-fusion-fission bombs reached yields of about 500 kilotons. In the 1950's Edward Teller & Stanislaw Ulam found a way to initiate a purely fusion explosion with an atomic explosion. A blanket of plastic was heated by a conventional fission explosion. The energy of the detonation blew off a plastic covering and compressed a core of lithium deuteride. The core reached the extreme conditions required for nuclear fusion. These bombs reached yields of over 50 megatons. A 50 megaton explosion will blow a hole about a half mile deep and mile wide.

It was believed that the tremendous power of nuclear fusion would soon be harnessed for the peaceful production of energy. Fusion reactors would burn a virtually limitless supply of heavy water (deuterium). These reactors do not run away. They intrinsically shut down upon failure. Their waste products are shorter lived and cannot be used to build a nuclear bomb. Fusion offered a panacea of power production. The most easily fused elements, deuterium and tritium, require a temperature of 40 million degrees to ignite. Temperatures of 150 million degrees are required for the commercial production of power. No sold material can contain this hot plasma. The first attempt to produce hot plasma employed a magnetic device called a **Z** pinch. A

Energy, Cold Fusion, and Antigravity

magnetic pinch compressed and heated plasma. The temperature at the pinch proved to be many orders of magnitude too low. Magnetic bottles called Stellarators were tried next. It was discovered that hot plasma is not easy to confine. Like a hot headed prisoner it become increasingly unstable and escaped from the best of the magnetic traps. The hot, unstable plasma quickly leaked from twists in the device. These Stellarators were not viable. The dream of nuclear fusion was moved 30 years into the future.

Hope came with the design of a better magnetic bottle; the Tokomak. It was invented in the 1950s by Soviet physicist Igor Tamm and Andrei Sakharov (the same men who invented the Soviet H-bomb). An international effort, involving 20 billion dollars, has constructed a massive Tokomak in Cadarache, France. This Tokomak has also encountered design problems with the stability of its hot plasma. The plasma tends to release its energy, like a flash bulb, in a burst of light. This effect is known as the Greenwald instability. Hope exists that ITER's problems can be resolved. It is expected to produce energy by 2025. ITER may already be obsolete because it does not employ the latest generation of ceramic superconductors. The days of cheap energy do not appear to be returning with any version of the Tokomak.

Huge lasers at the National Ignition Facility in Stanford are being employed to trigger nuclear fusion. Small pellets, of about the size of the aspirin, are heated to immense temperatures by these lasers. The pellets are compressed (imploded), by the heat, to a density of 100 times that of lead. Temperatures of 100 million degrees Kelvin have been

Energy, Cold Fusion, and Antigravity

reached. These temperatures and pressures have been maintained for a picosecond. The system almost meets Lawson's criteria. The powerful lasers, used in these tests, require hours to reset. To reach commercial viability, the process must run at a rate of 10 implosions per second. The director of the Laser Initialed Fusion Energy System (LIFE), Dr. Edward Moses, states that this goal can be accomplished within ten years. In 2012 The LIFE facility encountered unexpected problems and the fuel pellets failed to ignite. Four billion dollars has already been invested in this project. The resolution of these problems is dependent upon the spending of a massive amount of additional money.

Jed Rothwell has commented that the introduction of a new technology often follows a major advance within an existing technology. The hot rare earth gas mantel emitted bright light when gas heat drove its electrons to high momentums. It was heralded as the savior of the town gas system. At about the same time generators were developed that could directly harness the magnetic component of the electrical force. The electric light appeared with this development. The electric light promptly dispatched the gas lamp; mantle too. An analogous circumstance exists within the field of hot fusion. Hot fusion emerged from a Newtonian understating of the momentum of moving nucleons. Technology is appearing that can directly harness the magnetic component of the strong-nuclear force. High momentum particles are not required. Cold Fusion has appeared with this development. Hot fusion will be promptly dispatched by an assortment of low cost lattice assisted nuclear technologies.

CHAPTER 2 The Genesis of Energy

Energy, Cold Fusion, and Antigravity

There universe contains an enormous quantity of energy. This energy is not where it is needed and it does not come when it is wanted. Capturing solar energy with orbiting solar arrays is prohibitively expensive. Nikola Tesla stated that someday we would be able to *"tap directly into nature's wheelworks"*. In order to tap into this wheelwork a new question needs to be asked. Where did the energy of the universe energy come from in the first place? An understanding of the genesis of energy may lead to the development of new terrestrial sources of energy.

The genesis of the universe has been a subject of study and speculation by the greatest minds in philosophy and science. The original ideas on genesis were produced by the Greek philosophers. It became apparent to the Greeks that all things came from other things. The Greek Empedocles (495-435 BC) announced that only fools (who have no far-reaching thoughts) would believe that something could come from nothing. The noted Greek philosopher Plato (427-347 BC) described the concept of a form. According to Plato the form was the property that made a thing what it was. Plato's student, Aristotle (384-322 BC), developed the idea of forms and concluded that each form was composed of a substance. The form of a substance could be changed but the substance itself remained eternal. Where did the original substance come from? The conclusion that the ancient Greeks drew was that a prime mover created the original substance. This prime mover was God.

In the Middle-Ages the greatest thinkers on the subject of creation were theologians. One of these theologians was St. Augustine (354-430 AD). It became apparent to St.

Energy, Cold Fusion, and Antigravity

Augustine, like it did to the Greeks, that all things came from other things. If the substance of the universe was not created, then this substance must have always been. If the substance of the universe was eternal, then time had no beginning. Every event is preceded by a prior event. Without a first event, everything that could have happened should have happened in the infinite past. For current time to have meaning, the chain of events leading to the events of today must have had a started at some point. St. Augustine described this starting point as the first event. The first event was the genesis of the universe. St. Augustine used the Latin word ex-nihilo to describe genesis. St. Augustine concluded that an infinite source or prime mover created the substance of the universe ex-nihilo. St. Augustine concluded, once again, that the source of stuff was God.

Energy, Cold Fusion, and Antigravity

THE CREATION

Today scientists study the process of creation. Science does not address the question of who created the universe. Scientists have rules by which they work by called the conservation laws. Some of these conservation laws are the conservation of momentum, energy, and charge. Science asks the question, "How could the universe have formed within the the possibilities lying within the conservation laws?" According to current theory and experimental evidence these conservation laws hold true, everywhere, and at all times. The scientific principle of the conservation of energy was established, in 1847, by Mayer and Von Helmholtz. It restates the old idea that something cannot come from nothing. This principle is also expressed within the first law of thermodynamics. According to the accepted theory of the big bang the universe sprang from nothing 13.5 billion years ago. How did this happen within the bounds of conservation laws?

In 1973, the distinguished contemporary scientist Edward P. Tryon published a paper, "Is the Universe a Vacuum Fluctuation?". Tryon explained how the universe could have materialized without violating the principle of the conservation of energy. Tryon theorized that the total energy of the universe is zero. He stated that the positive energy of the universe's mass is balanced by a negative gravitational potential. It is well known that when something falls it loses gravitational potential energy. Take, for example, the energy produced by falling water at a hydro-electric power plant. The energy produced by the falling water is equal to the

Energy, Cold Fusion, and Antigravity

product of the weight of the falling water and the distance through which it falls. The relationship between energy, force, and distance is presented below.

$$Energy = Force \times Distance$$

According to Tryon, if a mass **M** were to fall to the edge of universe from an infinite distance away, the gravitational potential energy lost by the mass would equal its mass energy. The mathematics used in this computation was applied on a grand scale; however, its formulation is the similar to the mathematics used in the simple water fall calculation. The universe was considered to be one large, round sphere. The force of gravity is not constant. It diminishes with the square of the distance from its source. Newton's invention of calculus is commonly used to compute the energy lost by a mass that is pulled by a force that varies with displacement. The left side of the equation below is Einstein's famous equation for mass energy. The right side of the equation gives the work produced as the variable gravitational force acts through a fixed distance. This exchange of energy was stated with calculus.

$$Mc^2 = GMM_u \int_{\infty}^{r_u} \frac{1}{r^2} dr$$

The limit of the calculation r_u states that matter's rest energy equals its gravitational potential at the edge of the universe. The equation contains one unknown the mass of the universe M_u. A solution produced the mass of the universe.

Energy, Cold Fusion, and Antigravity

$$M_u = 2 \times 10^{53} \text{ kg}$$

This amount of mass will set the energy of the universe to zero. This is the quantity of mass required to conserve energy upon the genesis.

This result was compared to the mass of the universe as calculated from its density and volume. The universe has expanded beyond the visible radius. This portion, of the universe, is hidden behind a gravitational discontinuity and it does not interact with the visible universe. The visible universe was considered to be a sphere with a radius of 13.5 billion light years. It was assumed that it is filled with matter of the same density as our local space. This density is about four tenths of a proton of ordinary matter and eight protons of "dark" matter per cubic meter.

$$M_u = \left(\frac{.4 \text{ protons}}{m^3} + \frac{8 \text{ protons}}{m^3} \right) \times Volume \times M_p$$

The mass of the universe produced by the product of its density and volume was given below.

$$M_u = 1.7 \times 10^{53} \text{ kg}$$

Energy, Cold Fusion, and Antigravity

The vastly different constructs produced a similar result. Amazingly, these results agree to within 15%. This agreement demonstrates the validity of Edward Tryon's hypothesis. This argument reveals that the principle of the conservation of energy must be reinterpreted. It must now allow for perpetual motion. The negative gravitational potential of the newly formed energy cancels its positive energy. This has occurred at least once and it may be possible again.

In the nineteenth century, Ernst Mach (1838-1916) proposed that inertia is an effect of the influence of distant matter of the universe. For example, spinning matter experiences a centripetal force. Mach maintained that fixed local matter would also experience a centripetal force if the universe started spinning around it. The argument commonly used to dispel this idea is the same argument that was used to dispel the genesis process. It states that distant objects cannot influence local events. Essentially this argument asserts that no influence can travel faster than the speed of light. Given that distant regions of the universe are billions of light years away, these regions cannot energetically couple to a local event.

The arguments based on time and distances are not valid. For the sake of argument let's first assume that forces propagate instantaneously. The resulting forces are equal and opposite. The movement of local matter, in such a system, immediately affects the distant regions of the universe. No additional forces, other than the original exchange forces, are required. These static fields would, conserve momentum through their own actions. In the real

Energy, Cold Fusion, and Antigravity

universe forces do not propagate instantaneously. It takes time for the gravitational field of matter to establish itself throughout space. Moving matter immediately experiences the force of the established gravitational field of distant matter. The fields of moving matter require a finite amount of time to propagate outward. For a period of time, distant matter will continue to be attracted to the moved matter's old position. During this period, the original static forces are not capable of conserving the system's momentum. There is some uncertainty in the measurement of momentum; however, momentum is exactly conserved at all times and at all scales. Additional forces are required. Michael Faraday discovered that a moving electrical charge induces a magnetic field. A magnetic field is produced by the dynamic movement of an electrical charge. Its magnitude is proportional to the charge's speed. When a moving electron passes through a changing electrical field, a second magnetic field is produced. The magnitude of this second field is proportional to the rate of change of the external electrical field. A local force is produced by the interaction of these two magnetic fields. The action of this local electro-magnetic force balances the momentum of the moving electrical charges.

Einstein's equations demonstrate that moving matter induces a gravitomagnetic field. The gravitomagnetic field has a structure similar to the electric-magnetic field; however, it is of a gravitational origin. The gravitomagnetic field affects mass not electrical charges. The gravitomagnetic field carries the inertia of a moving matter. In a system consisting of moving force fields, momentum is conserved through the induction of transient magnetic forces (magnetic, gravitomagnetic, and strong nuclear magnetic). Faraday expected that a static magnetic field would impose a continuous electrical potential. He discovered that only varying magnetic fields induced an electrical potential. Faraday did not know why the process of electromagnetic induction was coupled with time. The element of time must be coupled with the process of induction in order to conserve momentum within a universe in which disturbances, in the static force fields, propagate at finite velocities. Gravitational linkages couple local events to remote regions of the universe. This coupling is accomplished through the introduction of dynamic gravitomagnetic forces. These magnetic forces act as a reservoir of momentum and energy. This reservoir stores energy and momentum until it can be transferred to distant regions of the universe. Local positive energy is linked to the negative gravitational potential of the universe through the interplay of transient gravitomagnetic interactions. In time, the books balance and the magnetic reservoir is depleted. In time energy will be, once again, conserved through the action of the original static force field.

The force fields conserve momentum through the process of induction. The behavior is exhibited by **all** of the natural forces and it is independent of scale. This action is a universal property of our universe.

Energy, Cold Fusion, and Antigravity

THE GENESIS PROCESS

Scientific arguments have shown that it is energetically possible to create substance from nothing. These arguments have tremendous philosophical implications. Can man now create something out of nothing? If he does he will acquire a power that was, since antiquity, considered to be exclusively within the domain of the Gods.

Inventors have been trying to build perpetual motion machines for centuries. The latest generation of inventors has been trying to extract zero point energy through a process of Casimir collapse. None of these inventions has worked and the patent office currently rejects all applications for patents on such machines.

"Joe Newman made one contribution to society in his lifetime, by suing the Patent Office for denying him a patent. The 1986 decision in Newman v. Quigg (the Patent Commissioner) is now cited as the authority for denying patent applications for perpetual motion machines out of hand." Dr. Robert Park, The American Physical Society

If something can be created out of nothing without violating the principle of the conservation of energy; what then inhibits the process of genesis? Does the process require exotic conditions like those that prevailed during the birth of the universe? Can genesis occur today and under ordinary conditions? What kind of a machine could make something from nothing? *"The best way to predict the future is to create it yourself!"* Peter Diamandis'.

Energy, Cold Fusion, and Antigravity

It has been known, since the time of Boltzmann, that energy flows from a concentrated low entropy state to a dispersed high entropy state. The heat in a cup of coffee, for example, tends to disperse throughout a room. The heat of a room, however, does not tend to concentrate within a cup of hot coffee. The tendency is expressed by the second law of thermodynamics. The dispersion is mathematically quantified in units of entropy. The entropy of the highly concentrated energy, at the moment of creation, must have been low. Sean Carroll has written,

"Why did the universe have low entropy near the Big Bang? It's not hard to understand why entropy increases; what's hard to understand is why it was ever low to begin with."[31]

Carroll's question can be restated. Is low entropy an essential ingredient of the genesis process? Can the low entropy condition of the original energy provide a clue to its origin? These questions suggest that a genesis machine may require the assistance of a low entropy superconductive Bose condensate.

The entropy of the original energy may have been zero. Energy, at zero entropy, is concentrated within a single body. The single bodied, low entropy, early universe spun without reference. The principle of the conservation of angular momentum had no meaning at this moment. Is the dismissal of the conservation of angular momentum an essential ingredient of the genesis process? A genesis machine must also produce new energy without regard to angular momentum. This action can be accomplished by producing two particles of energy; each with an opposite spin. The

Energy, Cold Fusion, and Antigravity

coupling of spins, with the materialization of energy, lies at the root of Einstein's spooky action at a distance. The contemporary universe has expanded beyond its visible extent. The positive energy of the existing stuff no longer couples to the negative gravitational potential of the matter beyond the visible horizon. This decoupling may have frozen the universe into the state we now enjoy. An isolated system will have to be constructed, in the lab, containing particles with masses that can couple with the current value of the negative gravitational potential. The vibrations in a Bose condensate are elementary and can carry variable amounts of energy. A condensate of protons may exist within the domains of a proton conductor. Does process of cold fusion progresses through the vibration of these protons. Can genesis occur within the superconductive regions of a proton conductor? Physicist Steven Jones argued that the process that is being called cold fusion is probably not nuclear. What is the magic that is producing the excess heat? Can energy materialize from nothing? Will the addition of energy upset the balance of the universe and make it uninhabitable?

The Heisenberg Uncertainty Principle has shown that vacuum fluctuations do generate anomalous spikes of energy. Many events, in the quantum domain, originate as a condition of these temporary bursts of energy. These bursts of energy quickly return to the vacuum. They cannot be employed to perform any useful work. The attempt, to harness this zero point energy, has failed.

Energy, Cold Fusion, and Antigravity

The hydroelectric, pictured above, converts the gravitational potential of dropping water into electrical energy. A genesis machine would harness the gravitational potential of a dropping universe.

This author strictly adheres to the principle of the conservation of energy. He suggests that these bursts of positive energy are accompanied by the negative potential of a strong, local gravitomagnetic field. This emergent strong gravitomagnetic field can be employed. The analysis of the path of the quantum transition, later in this text, reveals that the quantum transition requires a burst of anomalous kinetic energy. This kinetic energy cannot be tapped; however, the strong local gravitomagnetic field that accompanies it can be harnessed with technology. This utilization will allow for a classical exploitation of all of the natural forces. Open, fresh minds must begin exploring this possibility.

Energy, Cold Fusion, and Antigravity

CHAPTER 3 Electromagnetic Induction

In ancient times (221 B.C.) a rock, called lodestone, was floated in a dish of water. The lodestone formed a natural compass. The magnetic influence, upon the stone, was mysterious. The magnetism reached through space unseen, touched the stone, and turned the compass. The magnetic influence was invisible, permeated all of space, and exerted a force on all of nature. In 1600, William Gilbert likened this action to hand of God. In the 1700's, Benjamin Franklin rubbed glass and amber rods with fur and cloth. A charge of static electricity built up upon the rods. When the rod was placed near a grounded object a spark of electricity leaped from the rod. The electrical force exerted other mysterious influence; it attracted and repelled dust particles, hair, and fine particles. The ancient Hindu scriptures describe the Brahman. It's the basis of the material world, a God like force that holds all things together, and the hidden power that is latent in all things. The existence of the Brahman could only be recognized by special people with a keen insight. The invisible magnetic and electrical forces reach through space and bind matter as would the Brahman.

In the 1820s, Hans Christian Orsted built a crude liquid battery in a glass jar and ran the current through a wire. He demonstrated the experiment before his class. He sat, by chance, a compass near the wire. The compass deflected as the current flowed through the wire. Orsted discovered that a flowing electrical current induced a magnetic field. Orsted showed that the mysterious forces of electricity and magnetism were intimately interconnected. Professor Walter Lewin of M.I.T. has stated,

"Orsted's discovery was momentous". It rates right up there with the discovery of fire, gunpowder, and the wheel."

This discovery was employed in the building of generators, motors, transformers, ignition coils, and speakers. The process of electromagnetic induction forms the foundation of a large part of our modern technology. It is fundamental to the electrification of our civilization.

The action revealed the beautiful symmetry through which the natural forces act. A face is seen as beautiful, for example, when the left and right sides of the face are symmetric. The electric and magnetic fields exhibit a beautiful symmetry in that a flowing electrical current induces a magnetic field and, symmetrically, a changing magnetic field induces an electrical current. In 1861, James Clerk Maxwell, one of the greatest geniuses all time, quantified this symmetrical relationship within his equations. The formulations of James Clerk Maxwell were fundamental, symmetric, and beautiful. They revealed that the action of creation was symmetrical. With this understanding; Maxwell appeared to have read the mind of God. This reading did not reveal a God who acted upon the concepts of good and evil. This God did not speak in any natural human language. Maxwell discovered that the actions of the creator revolved around a beautiful symmetry. These actions were precisely spoken in the unnatural language of mathematics. Maxell's equation was presented next.

Energy, Cold Fusion, and Antigravity

$$\oint E \cdot ds = -\frac{d}{dt} \int B \cdot da$$

Both sides of Maxwell's equation are mathematical formulations called integrals. This left side describes the electrical voltage E around a closed loop of wire. The right side describes the rate of change of the magnetic field B through that loop. In English the equation states that the voltage around a closed loop of wire is proportional to the rate of change of the magnetic flux through that loop; and vice versa. This equation precisely described the process of electromagnetic induction. This process was built into the technology that currently immerses us. The electric and the magnetic forces bind the substance of the universe. They act like the mortar of a brick wall. Their action, as Gilbert has suggested, is like that of the hand of God.

An electrical current carries an impetus. This impetus opposes a change in an electrical current. An automotive spark coil employs electrical impetus to its advantage. The ignition system quickly interrupts the current through a coil. The impetus of the electrical current, within the spark coil, presses forward through a high voltage arc. James Clerk Maxell showed that the impetus of an electrical flow produces a voltage in opposition to a change☐ in current.

Voltage = Inductance x ΔCurrent

Similarly, Isaac Newton showed that the matter produces a force in opposition to a change in its velocity.

Energy, Cold Fusion, and Antigravity

Force = Mass x ΔVelocity

These two equations have the same form. This form shows that electrical voltage is analogous to mechanical force and that electrical inductance is analogous to inertial mass. The essence of inertial mass is unknown; however, the constituents of an electrical inductor are well known. A ferromagnetic core can be placed within an electrical solenoid. The soft iron in the core soaks up the magnetic flux like a sponge. Electrons orbit the nucleus in ferromagnetic iron. These electrons form tiny current loops. A small magnetic field is produced by the residual of this spinning motion. The magnetic fields, of billions of atoms, line up in unison within a grain of ferromagnetic material. The magnetic fields of the grains, however, are randomly oriented. The magnetic field of the grains can be lined up through the application of an external magnetic field. That's how a permanent magnet is made. Magnetism is not a conserved property and the insertion of ferromagnetic core into an energized solenoid can increase its magnetic field by a factor of 10,000 or more. The additional magnetic flux emerges from nothing, ex-nihilo. The additional magnetic flux increases the inductance of an electrical solenoid. Given that electrical inductance is analogous to inertial mass there should be conditions that increase inertial mass. Special Relativity has revealed that the addition of velocity is one such condition.

Vibration destroys the composite magnetic field of a ferromagnetic material. If an electrical pick up coil placed near a permanent magnet as the magnet's atoms are vibrated

Energy, Cold Fusion, and Antigravity

by heating, it will detect the randomization of the orientation of the grains. If the output of this coil is fed through an audio amplifier clicking sounds can be heard. The clicks are much like those of a Geiger counter. The coherer is an early radio receiver (circa 1910) that employed this phenomenon. It produces a clicking sound as the magnetic field of a ferromagnetic metal powder was destroyed by radio frequency vibration. Magnetic amplifiers also work through a process that effectively destroys the inductance of a ferromagnetic core. These composite magnetic fields do not go somewhere else. They are not conserved and just go away. Given that electrical inductance is analogous to inertial mass, are there conditions where vibration destroys inertial mass?

I recall being in Professor Richard Bender's class at The University of Pittsburgh at Johnstown in 1973. Professor Bender was going over the air core transformer. The simplest embodiment of an air core transformer is two inductors connected in series. Professor Bender stated that the inductance of the combination was equal to the self-inductance of the first inductor plus the self-inductance of the second inductor. This made good sense to me as the physical construction and the current within the inductors did not change. Then he added on an additional exponential term that accounted for the mutual inductance of the two interacting inductors. I thought, *"Where did that come from? Why does everything have to be so hard? Surely, I am going to flunk!"* I realized, much later in life, that magnetism is not a conserved property of the universe and that the mutual flux emerged from nothing, ex-nihilo. Professor Bender moved

Energy, Cold Fusion, and Antigravity

on in his lecture and I remained stuck right there. The renowned professor Walter Lewin of MIT has stated that magnetism is non-conservative but I have never heard him specifically state that magnetism is not a conserved property of the universe. So I will say it. The system that the magnetism emerges from must obey the conservation laws; however, the non-conserved properties of magnetism, and by analogy inertial mass, are beyond mathematics. Newton grasped this as he wrote his equations based upon the conservation of momentum, not inertial mass. Magnetic fields can appear and disappear at will. They can be stretched to galactic distances. The spin of an entanglement states may be connected by this action. Later in this text it will be shown that the quantum condition emerges from classical Newtonian mechanics as a consequence of the non-conservation inertial mass. This non-conservation has gone largely unnoticed and this text will go on show that it is of the upmost importance.

The static electrical field is conserved. Dielectrics can only decrease the strength of the static electric field. A superimposed field of the opposite polarity, within a dielectric, reduces the strength of the composite static field. No material or condition can increase the strength of any static field. I know of only two things that are not conserved; magnetism and internal mass. A strong composite magnetic field is produced within paramagnetic materials. Soft iron is used, within electrical equipment, to greatly amplify the strength of the electro-magnetic field. This strong magnetic field is employed, through a process of induction, to power our economy.

Energy, Cold Fusion, and Antigravity

GRAVITATIONAL INDUCTION

The dynamic electro-magnetic field is not conserved. The magnetic field emerges, as needed, to conserve the momentum within a system of moving static force fields. No disturbance can propagate faster than light speed and all of the natural forces conserve momentum through the process of induction. The induced component of the gravitomagnetic field, as expressed by General Relativity, is given below.

$$B_g = \frac{K_g G}{c^2} \frac{dm}{dt}$$

General Relativity revealed that moving mass (**dm/dt**) induces a gravitomagnetic field in the same way that a flowing current (**dq/dt**) induces a magnetic field. Einstein called the effect frame dragging. The induced gravitomagnetic field has the same curly structure as the electro-magnetic field; however, it is not of electromagnetic origin. The gravitomagnetic field carries the momentum of moving matter. Its strength is limited by the low permeability of free space to the gravitomagnetic field, **G** divided by **c** squared. The gravitomagnetic field is ten to the thirty ninth power weaker than the electro-magnetic field. It is vanishingly weak. The gravitomagnetic field, produced by the rotation of the entire earth, is so weak that it was barely detectable by the most sensitive orbiting gyroscopes.

Energy, Cold Fusion, and Antigravity

The relative permeability expresses the amplification of the magnetic field within a paramagnetic material. The relative permeability of the gravitational field $\mathbf{K_g}$ was always considered to be one. It was never factored into the equation. The strength of the gravitational field is considered to be fixed and ten to the thirty ninth power weaker than the electromagnetic field. The Earth's gravitational field is easily overpowered with a tiny magnet. It takes the whole of the earth to produce the gravitational field that we experience. The Earth's mass cannot be spun within a machine in order to harness gravitomagnetic induction. This author suggests that the strength of the gravitomagnetic field, as with all of the dynamic magnetic fields, is not a conserved property. Murray Gellman has stated, *"Anything that is not forbidden* (by the conservation laws) *is mandatory."* Soft iron dramatically increases the strength of the electro-magnetic field. There should be a material or a condition that affects the strength of an induced gravitomagnetic field. This material would have a large permeability $\mathbf{K_g}$ with respect to the gravitomagnetic field. The lining up of electronic spins within ferromagnetic material provides a clue as where to look for a para-gravitomagnetic material. A para-gravitomagnetic material would strengthen the gravitomagnetic field. Could the alignment of the nuclear spins amplify the gravitomagnetic field and produce a para-gravitomagnetic effect? Nucleuses line up in the presence of a strong magnetic field. An MRI imaging machine, for example, works through a process that depends upon an alignment of the nuclear spins. No gravitomagnetic anomalies have been reported emanating from MRIs or strong neodymium magnets.

Energy, Cold Fusion, and Antigravity

In 1995, Scientist Dr. Ning Li suggested that nucleuses that are both aligned and spin up may induce a significant para-gravitomagnetic effect. She was not able to demonstrate such a technology. In 2013, Valery Milner, a molecular physicist at the University of British Columbia kicked molecules of a gas with a powerful 100-trillionths-of-a-second laser pulse. The molecules were spun up to 10 trillion rotations per second. Milner stated, *"There's definitely nothing macroscopic that can spin that fast. A car with tires turning at that spin rate would travel the distance to the nearest star in half an hour."* Is it possible to induce a strong gravitomagnetic field with this technology?

Superconductors are a new state of matter. This state profoundly affects the electro-magnetic field. Superconductors are dia-magnetic and completely impermeable to a magnetic field ($u_o=0$). The magnetic field inside the superconductor is zero. Magnetic flux is driven out beyond the surface of a superconductor. This action is known as the Meissner effect. It is the opposite of a flux increasing paramagnetic effect. The effect binds the magnetic dipoles of trillions of atoms into a new state of matter. A strong composite magnetic field rests upon the surface of a superconductor. Maglev trains are lifted by this strong electro-magnetic effect. The static gravitational field has no field of the opposite polarity. It cannot be reduced by shielding Dia-gravitomagnetic materials are not; however, beyond the realm of possibility. These dia-gravitomagnetic materials would expel the gravitomagnetic field of trillions of spinning nucleons to their surface. The expelling force, as shown later in this text, is strong and it may be harnessed.

Energy, Cold Fusion, and Antigravity

Frank Znidarsic and Dr. Hal Puthoff in 1992

In his quest to find a dia-gravitomagnetic material Frank Znidarsic traveled, in 1992, to the Institute of Advance Study in Austin. He spoke there with Dr. Hal Puthoff. At that time Puthoff was working with a concept called a Goldstone Boson. This Boson purportedly had strong gravitomagnetic effect.

No strong gravitomagnetic effect was detected emanating from any superconductor. Superconductors were frozen up, at rest, an appeared to do nothing. What are the special ingredients needed to manufacture a dia-gravitomagnetic material?

Energy, Cold Fusion, and Antigravity

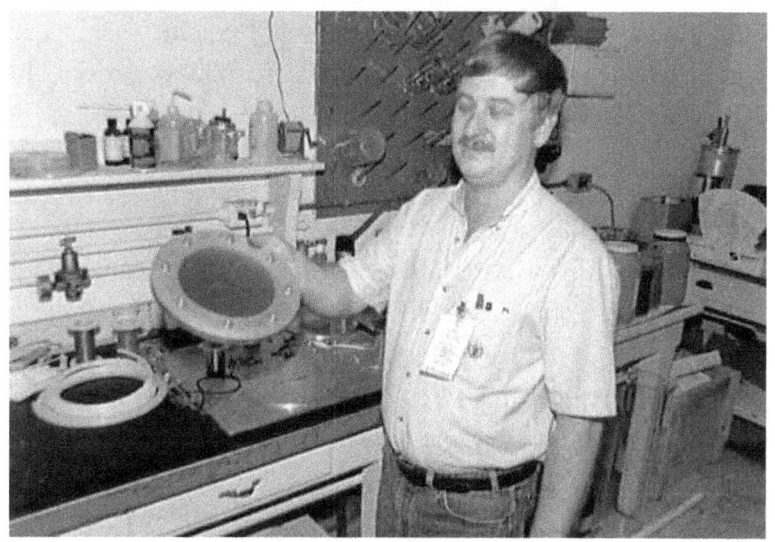

NASA scientist Glen Robertson in 1998

In 1998 Frank Znidarsic visited the Marshall Space Flight Center in Huntsville Alabama. They were conducting gravitational experiments based on the work of Yogini Podkletnov. Podkletnov vibrated a layered, spinning superconductive disk with radio waves. He reported that a strong gravitational anomaly was produced when the magnetic field was operating at high frequencies, on the order of 3.2 to 3.8 MHz.[45]. Do vibrations expel the gravitomagnetic field of a rotating superconductor? Does this action send the superconductor into a state of macroscopic quantum transition?

NASA made superconductive disks and spun them. Glen Robertson is holding a magnetic disk used in NASA's experiments. NASA's budget was cut, by short sighted

Energy, Cold Fusion, and Antigravity

individuals, and they never applied radio frequency stimulation or added layers to their superconductive disks. In 2011, Glen Robertson and Frank Znidarsic went on to publish a peer reviewed scientific paper "The Flow of Energy" in "Physics Procedia".

In a parallel effort BAE systems attempted to replicate Podkletnov's results in the UK. They also assumed that the only purpose of the radio frequency stimulation was to rotate the disk. Ronald Evans an engineer with BAE system's project Green Glow wrote.

"Limited funding meant that Podkletnov 's method of using Meissner levitation and an interaction with a radio frequency field to generate the rotation of the YBCO disc was not attempted...However, this still leaves the possibility that it was Podkletnov's method of electromagnetic excitation of his superconducting annular disc that gave rise to a gravitational effect."

Podkletnov applied a three megahertz radio wave to a 1/3 of a meter disk. The product of the frequency and diameter equaled one million meters per second. Znidarsic suspected that this speed S_n was fundamental to the induction of a strong gravitomagnetic field but, at the time, he had no clue as to how or why. David Noever of NASA introduced me to their related downshifting the frequencies theory. Could the process of vibration profoundly influence the relative gravitational permeability K_g of a superconductor? Is it possible to induce a strong (ten to the thirty-ninth power stronger) gravitomagnetic field? Why is the velocity of one million meters per second so special?

Energy, Cold Fusion, and Antigravity

Empirical COLD FUSION

In a parallel development Martin Fleishmann and Stanley Pons claimed to have discovered that fusion progresses within a proton conductor. Pons and Fleishmann were renowned scientists. There announcement in 1989 produced a firestorm of activity. Dr. Robert Park (spokesperson for the American Physical Society) John R. Huizenga (a renowned nuclear scientist at the University of Illinois), and Dr. Richard Garwin (a renowned nuclear bomb scientist) jumped into the fray. They 'knew' from well-established theories that room temperature fusion was not possible. Fusion requires a minimum temperature of 40 million degrees K. The lack of radiation and emitted neutrons completely ruled out the possibility of any type of nuclear reaction. The cold fusion process proved to be fickle and was not reproducible. The scientific establishment turned its back on cold fusion. The magazine <u>Scientific American</u> refused to publish articles on cold fusion. The patent office automatically denied all patents based upon cold fusion. This story is one of the greatest fiascoes since the clergy refused to look through Galileo's telescope. A small cadre of independent inventors continued to generate heat within their cold fusion cells. In 1990, Dr. Robert M. Bass published a paper, "<u>Stimulation / Actuation, and Automatic Feedback Control of Cold Fusion</u>". This paper suggested that thermal vibrations expedited the cold fusion process. Are the thermal vibrations in a cold fusion cell related to the radio frequency vibrations within Podkletnov's device?

Energy, Cold Fusion, and Antigravity

**Frank Znidarsic and James Patterson
December 1995**

James Patterson was one of the first to produce large amounts of heat. Znidarsic traveled to Anaheim California took a peek at Patterson's cell. By chance Dr. George Miley, a great contemporary scientist, also attended the gathering. Patterson used a preheater to get the reaction started. How can low energy thermal vibrations influence a high energy nuclear reaction?

"It may be that once excess heat is initiated that the nuclear energy goes into the vibrations maintaining the excess heat process." Peter Hagelstein MIT [48]

It was also becoming apparent that cold fusion progressed within a nanometer domain. Jed Rothwell wrote in "Infinite

Energy, Cold Fusion, and Antigravity

Energy, Issue 29, 1999, page 23", "*50 nanometers ...is the magic domain that produces a detectable cold fusion reaction*". Electrical conductors convey electrons and proton conductors convey protons. Hydrogen nucleons flow through a proton conductor in reaction to an applied voltage. Palladium deuteride and nickel hydride are proton conductors. In 1971, Cameron Satterthwaite, professor emeritus of physics, was the first to discover superconductivity in palladium deuteride. Many others have since confirmed this discovery. This superconductivity exists, within nano-domains, at room temperature. Can nucleons, dissolved within a proton conductor, bind into a single collective state? A collective state of protons is known as an inverse Bose Einstein condensate.

It took a while to put it together but Znidarsic finally figured it out. The speed of a traveling wave is normally expressed in units of meters per second. The speed of a standing wave is zero. The vibration of a standing wave can be expressed as the product of its frequency (in cycles per second) and its wavelength (in meters). Units in hertz-meters are equivalent to units in meters per second. The apparent speed of standing wave can be expressed in hertz-meters. The product of the thermal frequency (20 terahertz) and the palladium grain size (about 50 nanometers) equaled an apparent speed of one million hertz-meters. Znidarsic had previously observed this speed within Podkletnov's superconductive disk. The radio waves and the thermal vibrations gave a superconductor a kick and sent it into motion. The vibration of a superconductor or a proton conductor, at a dimensional frequency of one million hertz-meters, reinforced the

Energy, Cold Fusion, and Antigravity

condensate, and formed a new state of matter[41]. This state is one of continuous quantum transition. Gravitomagnetic flux was diamagnetically expelled from the transitional state. Was the expelled gravitomagnetic field producing power through a process of genesis?

Dr. George. M. Miley completed an analysis, of Patterson's cell, at the University of Illinois. In 1997, Miley presented this result at the Wright Patterson air force base. Znidarsic attended the meeting. Miley plotted the products of the reaction on a periodic chart. The result was U shaped curve. The fracture of a heavy element produces a distribution of daughter elements. These daughter elements lie on a U shaped curve. The original heavy element rests at the bottom of the U. Miley interpreted the distribution of daughter elements to be the result of the fission of a heavy, collective nucleus. This collective nucleus had an atomic mass of 154. It was almost twice as heavy as Uranium! It appeared that a collective nucleus magically emerged within a cold fusion cell. The process of cold fusion was subsequently renamed and it is now known as a low energy nuclear reaction.

The interior of the sun is just hot enough to slowly fuse the heavier elements. The transmutation of elements heavier than iron requires the tepid energies of a supernova explosion. Existing theory does not allow for the emergence of a multi-bodied collective nucleus under any circumstance. Miley's results were widely criticized. When Miley claimed that the heavy elements had transmuted at room temperature, some asked if Miley had gone mad. It was said that Miley had fooled himself and that his experiment was contaminated. Miley countered that he had detected

Energy, Cold Fusion, and Antigravity

elements with abnormal isotropic ratios (different numbers of neutrons). These ratios did not exist in nature and could only be the result of a nuclear reaction. Paterson subsequently encountered trouble with the reproduction of his results. The lack of reproducibility, once again, provided more ammunition for the antagonists. Slowly and surely other independent scientists also began to produce results. It appeared that the diamagnetic effect extends to the magnetic component of the strong nuclear force. This magnetic component of the strong nuclear force is known as the spin-orbit force. The spin-orbit force accounts for the stability of the even numbered elements. Did a universal diamagnetic effect expel the nuclear forces to the surface of a nano-particle? As demonstrated by the electromagnetic example such an effect lies within the bounds of the conservation laws. It is possible. The expelled nuclear forces would act upon a cluster of nucleons from the surface of the nano-particle. At the invitation of Dr. George Miley, in 2000, Znidarsic first presented his theorem at a meeting of the American Nuclear Society. His theorem states,

"The constants of the motion tend toward the electromagnetic in a Bose condensate that is vibrated at a dimensional frequency of 1.094 megahertz-meters".

This theorem states, in a scientific way, that the vibration of a proton conductor or a superconductor adjoins electrons and protons into a single collective state. The dynamic components of the force fields (electromagnetic, gravitomagnetic, and nuclear magnetic) are expelled to the surface of a nano-crystal. Dr. Miley made his crystals as large as possible. Miley quickly follow suit, and is now

Energy, Cold Fusion, and Antigravity

using nanometer sized crystals. The verdict is out. Many good scientists, such as Dr. Edward Storms, believe that the nano-grain size only facilitates the absorption of hydrogen gas. An admixture of lithium, in the form of lithium hydride, appears to facilitate the formation of the nano-meter sized grains.

The competing poly-neutron theory of Larson and Widom has recently gained some acceptance. This theory cannot explain the lack of emitted radiation. Nucleons overcome their mutual electrostatic repulsion by flying over the electrostatic potential barrier. The process is like driving over a speed bump. A thump is emitted as you hit the bump. A gamma ray and a neutrino are emitted, as a thump, as an electron flies over the electrostatic potential barrier of the nucleus. Theories that are based upon an electron screening effect are flawed. They cannot account for the missing radiation. A conventional nuclear reaction emits high energy gamma radiation. This radiation is extremely penetrating. Theories based on the absorption or scattering of this radiation are wrong. If these mechanisms efficiently absorbed gamma radiation they would already be employed within the shielding of conventional nuclear reactors.

Dr. George Mathos, of the University of Indiana, has stated that the only way in which a nuclear reaction can proceed without producing high energy radiation, is under a condition where the range of the nuclear forces exceeds that of the electromagnetic force. This action is akin to stepping over a speed bump. He also pointed out that if the range of the strong nuclear force is extended matter would be crushed out of existence. The static forces are conserved. The range of the static nuclear forces cannot be extended. The range and strength of the dynamic nuclear magnetic forces are not conserved. The induced magnetic component of the strong nuclear force is the spin-orbit force. The spin-orbit force has the same curly structure as the magnetic electrical field; however; it is of a nuclear, not of an electromagnetic, origin. It carries the impetus of the strong nuclear force. It, like the magnetic component of the electrical force, is ejected from a vibrating Bose condensate. The probability of a nuclear interaction is directly proportional to the amplitude of the vibration at the dimension frequency of S_n squared. The LENR reaction proceeds at long range when the collective vibration traverses a cluster of nucleons. This range exceeds that of the nuclear electrostatic potential barrier. No high energy radiation is produced by the smoothly acting process.

This activity is not a sideshow confined to an unusual condition within a few odd experiments. Its appearance addresses some long standing fundamental issues. The condition is that of a quantum particle. The action is that of quantum transition. The effect can be extended, through engineering, to trillions of atoms. These atoms can be forced into a single, macroscopic transitional quantum state. Long range nuclear magnetic forces and strong gravitomagnetic fields will be induced. This idea is on par with the discovery of Hans Christian Orsted. It shows a way to classically control all of the natural forces. The technologies that follow can provide an unlimited source of clean energy and they may propel us to the stars.

Energy, Cold Fusion, and Antigravity

CHAPTER 4 Spacecraft Propulsion

In 1798, Thomas Robert Malthus published "An Essay on the Principle of Population". Malthus wrote that population will always increase in a geometric progression. This growth will quickly outrun the food supply and lead to famine, war, and ill health. He suggested that the poor should not be helped. This relief would only lead to the growth of a more excessive population. As Britain began a period of global colonization this philosophy was applied to the Irish. As predicted, in the short run, a famine ensued in Ireland. The Malthusian philosophy has, however, proved not to hold in the longer run. New lands were colonized and new methods of farming were developed. In 1968, Stanford Professor Paul R. Ehrlich followed through on the ideas of Malthus in his book "The Population Bomb". Ehrlich predicted that the population would overtake the food supply by the late 1970s. A mass starvation would ensue. Ehrlich's predictions seemed to be more plausible than those of Malthus. In the 1970s there were no new lands available on which to offload the excess population. Newly engineered crops and energy intensive methods of farming held off Ehrlich's catastrophe. An unexpected drop in the birthrate also helped to fend off the impending disaster.

In the 1960s many scientists were invited to predict what the world would look like at the end of the 20th Century. Environmental scientist James Ephraim Lovelock made his prediction. In the end, Lovelock was the only one who got it right. He earned considerable credibility as a result of the accuracy of his prediction. He has now stated that population

Energy, Cold Fusion, and Antigravity

growth, along with an energy intensive lifestyle, have now defeated the earth's capacity to control its chemical and physical environment. He claims, in the second half of the 21st Century that the only inhabitable places on Earth will be at the poles. He has suggested that we should, *"Enjoy life while you can!"* Australian researcher Graham Turner has recently taken another look at this issue within a MIT funded study. Turner has also found that the world is on track for a disaster. He has concluded that world will run out of natural resources. Oil will peak by 2030 and the demand for food and services will rise. The energy intensive farming and food supply system will fail.

In the early 1900s it was believed that the next frontier would be space. Future populations would live in huge spinning space wheels. The rain forests of Venus would be a tropical paradise. Canals could be built, if they were not already there, to bring water from the poles of Mars to its equatorial regions. These projections were wrong. Space travel proved to be extraordinarily difficult. So difficult that, at this moment, the United States has no manned space program.

The first probes to Venus measured its surface temperate and its atmospheric pressure. The surface of Venus roasts at 900 degrees Fahrenheit. That's hot enough to melt lead. The pressure on the surface is ninety atmospheres. Every now and then the whole surface of Venus turns over in a massive volcanic eruption. Venus is no paradise; it is hell. Why was Venus so unexpectedly hot? Comparative planetology was now undertaken. Scientists discovered that a massive concentration of carbon dioxide in the atmosphere of Venus trapped a large amount of heat. This process became known

Energy, Cold Fusion, and Antigravity

as a runaway greenhouse effect. Man's use of fossil fuels has increased the carbon dioxide content of the Earth's atmosphere. Could a runaway greenhouse effect heat up and kill the Earth?

The first space probes to Mars revealed a cold and a cratered, moon like surface. The lack of carbon dioxide, in the atmosphere, cooled Mars more than expected. Mars is so cold that carbon dioxide solidifies in the winter. As a child I touched dry ice (frozen carbon dioxide). I quickly discovered that the dry ice was so cold that it burned. For a long list of reasons including the lack of a magnetic field, a thin atmosphere, freezing temperatures, and the presence of carcinogenic hexavalent chromium; Mars is not hospitable to life as we know it. It appears that the only home we have, in this solar system, is the Earth. There are a billion-trillion starts in the universe. A multitude of planets are now being detected around these remote stars. It has been projected that other habitable planets will soon be discovered.

Rockets must lift the weight of their reaction mass. The Tsiolkovsky rocket equation shows that the payload of the best rockets is limited to a few percent of this weight. Rockets reach orbit only by shedding pieces of the rocket through a process of staging. Rocket powered manned missions to Mars will be expensive, long, and dangerous. Rocket technology falls far short at interstellar distances.

In the 1950s nuclear space propulsion was being investigated. Concepts, such as a space born reactor that ejected a stream of fast neutrons, held promise. Such technology could propel a probe to the nearest stars within a

Energy, Cold Fusion, and Antigravity

lifetime. The launch of a nuclear reactor was wrought with danger. The time frame of this technology was still too long to take people to the stars. Physicist Stanton Friedman, has stated, that a propulsion system that could produce an acceleration of one **g** for one year could transit to the nearest stars within a few years. This kind of technology could take people to the stars; however, it is several orders of magnitude more powerful than any known technology.

A magnetic spaceship that could propel off of the earth's poles has been often proposed. Frank Scully wrote in his 1950 book, Behind The Flying Saucers;

"Now that we have learned that these ships from another world fly magnetically, it is my opinion that our own magnetic engineers will solve the problem of magnetic flight and match the ships of the visitors with saucers made on the earth."

Magnets do push off of each other for a short length; however, the long range force that acts upon a magnetic dipole is purely rotational. A magnocraft produces no translational force. It would just flip over.

There was a great hope that the unification the gravitational and electromagnetic forces would provide an answer. Donald Keyhoe wrote in his 1950 book, The Flying Saucers are Real;

"If gravitation were a manifestation of the electromagnetic force, would it be possible that an advanced race had found a way - as unique as splitting the atom - to offset gravity and utilize that force."

Energy, Cold Fusion, and Antigravity

It's been over 100 years since Einstein attempted to find a solution that could unify the natural forces. To date, no simple solution has emerged. It is now believed that a unification of the natural forces will occur at energy levels that will always be out of the reach of man's technology.

The strength of the magnetic action of a solenoid is increased, by a factor of 10,000, upon the insertion of a soft iron armature. This flux amplification does not require a unification of the natural forces. In the early 1800's, iron core electromagnetic solenoids were inserted into the nascent electrical technologies. Telegraphs, generators, and motors emerged as a result of this application.

Given the reality of this electromagnetic effect, there should be a material or a condition that increases the strength of an induced gravitomagnetic field. Podkletnov has reportedly demonstrated a rotating, vibrating superconductive disk that appeared to produce a strong gravitomagnetic anomaly. An analysis of the quantum transition, later in this text, quantifies this reaction. The induced gravitomagnetic anomaly is not of electromagnetic origin. The effect, like its electromagnetic analog, does not require a unification of the natural forces. The action emerges at very low cryogenic energies. The flux amplification is 10 to the 39 power beyond that of the expected. The strong, induced gravitomagnetic field should react di-magnetically and repel matter. This field may react with the atmosphere and efficiently lift a heavy spaceship into a near earth orbit. In the vacuum of space there is no air to react with. The universe consists mostly of dark matter and, as far as we know, it is evenly distributed throughout the galaxy. Dark

Energy, Cold Fusion, and Antigravity

matter reacts gravitationally and it may provide a universal source of reaction mass. It may also be possible to directly interact with dark matter at another dimensional frequency. These applications may allow man to efficiently reach orbit and to travel to the nearby stars.

An induced, strong gravitomagnetic field may allow a spacecraft to be propelled without the ejection of reaction mass. Einstein's principle of equivalence states that the inertial and the gravitational mass of matter are inseparable. The inertial mass of matter must be reduced along with its weight. For example, take the case of the electrical transformer. Alternating currents in the primary of the transformer are limited by the inertial impetus of the primary's electrons. This impetus is carried by the electromagnetic field. The electrical impetus is directly proportional to the inductance of the primary circuit. Currents in the secondary winding can reduce the flux in the core. This action reduces the inductance of the primary circuit.

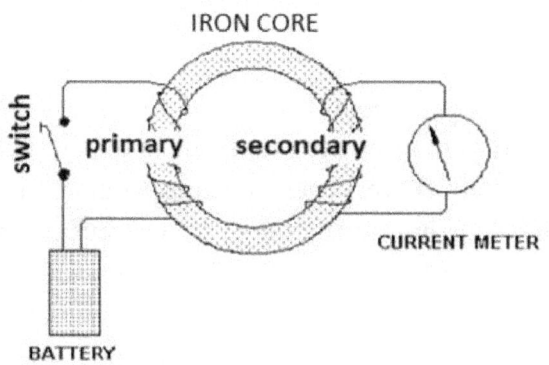

The Electrical Transformer

Energy, Cold Fusion, and Antigravity

The reduced inductance decreases the electrical impetus of the primary's electrons. The electrical current, in the primary, increases in reaction to the reduced impetus.

The inertial mass of matter is physically equivalent to the impetus of an electrical circuit. The gravitomagnetic field carries the physical momentum of a moving mass. Physical momentum appears as the universe reacts to an induced gravitomagnetic field. Special Relativity has shown that inertial mass can be increased through the application of speed. Internal mass is a dependent property and it is not, in itself, conserved. Its reduction lies within the conservation laws. Remote regions of the universe react to the composite gravitomagnetic field. The traveling matter's momentum will be conserved with the reduction of its inertial mass by an increase in speed. An analysis of the quantum transition, later in this text, quantifies the energy and the direction of this action. The reduction in the inertia of the spaceship will also make existing propulsion systems more effective. The reduction of the inertial mass of a surrounding medium may allow a vehicle to transit through it without emitting a sonic boom.

Travel at galactic distances requires superluminal speeds. Contemporary proposals for superluminal travel, such as the Alcubierre Warp Drive, depend upon the compression of space time. Starships would travel through this compressed space time at sub-luminal speeds. The compression of space time requires a tremendous amount of energy. Superluminal speeds appear to disrupt the sequence of cause and effect. They also, according to Special Relativity, require an infinite

Energy, Cold Fusion, and Antigravity

amount of energy. Superluminal speeds are considered to be inaccessible.

"The chronology protection conjecture" prohibits superluminal travel and communication.

Stephen Hawking

It may be possible to reduce inertial mass to zero. Just as an electromagnetic coil expels a diamagnetic material, the universe would tend to eject matter with no inertial mass. This material would lose mass energy as it accelerates and it would not be subject to the same restrictions as ordinary matter. A small piece of it tightly bolted down within a space ship could provide a propulsive source. A ship made entirely of it could leak out of the universe and pop back within another galaxy. Time paradoxes might be avoided because this condition is beyond the bounds of Special Relativity. The idea of something popping in and out of the universe is not without precedent. Entangled quantum systems appear to react at superluminal speeds. This communication may progress by the through the action of no inertial mass. The action provides a useful analog. Is it possible, by such a mechanism, to travel at superluminal speeds without a violation of causality? The development of these advanced applications may allow for unlimited travel.

It appears that a strong gravitomagnetic field can be induced with a low energy technology. This field needs to be produced and studied. The emergence of dynamic quantum effects, on a macroscopic scale, may result in some surprising applications.

Energy, Cold Fusion, and Antigravity

CHAPTER 5 The Quantum Mystery

Joseph von Fraunhofer invented the first spectrometer, in the early 1800's. He saw spectral lines within the light of the sun with this device. He used these lines, as reference points, in the design of achromatic lenses. In the mid-1800s, Robert Bunsen and Gustav Kirchhoff heated elements to incandescence within their Bunsen burner. They also saw spectral lines in the light of the hot flame. Johann Balmer produced an empirical equation that described this spectrum. Later in the 1800s, Johannes Rydberg extended Baumer's formulation to the spectra of the heaver elements. Astronomers employed spectral analysis and determined the elemental composition of stellar objects. The early scientists could not, try as they might, produce a causative explanation for the spectral emissions.

In the early 1900s, Max Planck showed that the intensity of the light emitted by a black body was an effect of the quantization of energy. Planck's abstract constant quantified this quantum of energy. Planck searched for and did not find a classical a classical explanation for this effect. Albert Einstein reused Planck's constant and showed that the energy of light is bundled into particle-like photons. Niels Bohr's principle of quantum correspondence appeared with this model. It states that the amplitude of a classical wave corresponds, in some mysterious way, to the frequency of a photon. According to the principle of quantum correspondence high frequency photons contain more energy just as large waves contain more energy.

Energy, Cold Fusion, and Antigravity

High frequency, low intensity ultraviolet photons can give a sun burn without heating. Alternatively high amplitude lower frequency visible light can heat without burning. Bohr's correspondence principle did not identify the root cause this duality and it is incomplete. An emitted photon exists as both a wave and a particle. These properties are mutually exclusive and their simultaneous emergence is a paradox. In an attempt to reconcile these difficulties Bohr introduced his second abstract principle; the principle of complementarity. This principle states that the frequency of quantum stuff varies linearly with the energy of a thing. The photoelectric effect is an expression of this duality. An uncertainty in quantum measurement looms within the shadows of this duality.

The electrical force varies exponentially. Why does the principle of complementarity act linearly? Why doesn't the quantization of energy emerge as an intrinsic property of the electric field? The ensuing quantum theory precisely described the quantization of energy; however, it did not explain the emergence of the quantum behavior.

Bohr applied Planck's construct to the atomic structure of the atom. Bohr showed that electrons jump between fixed orbits within the atom. The angular momentum, of the atomic orbits, was described a reduced (divided by 2π) Planck's constant. Bohr's, solar system like, semi-classical model quantified the emission spectrum of the atoms and determined the chemical properties of the elements. The atomic orbits reside at integer squares of the ground state radius. Classical harmonics appear at integer multiples (not squares) of the frequency of the motion. Why are the atomic orbits not spaced harmonically? It was said for this, and other reasons, that the atomic action was not that of classical harmonic motion. It was assumed that the motion emerged as a consequence of the mysterious, non-classical property of quantum spin.

According to classical electromagnetic theory of James Clerk Maxwell orbiting electrons should continuously emit electromagnetic energy. Atoms emit packets of energy at random intervals. Bohr's model cannot explain the stability of the stationary atomic states, produce the probability of transition, or explain why the frequency of the emitted photon is not coupled to the frequency of the orbiting electron. The frequency of the sound emitted by a speaker, for example, is coupled to the vibration of its cone. No single atomic orbit, or a combination of orbits, vibrates at the frequency of the emitted photon!

The contemporary workaround presumes that the quantum realm is the true reality and that the everyday classical domain appears as a subset within this larger reality. How can the domain of the whole be less than that of its parts? Classical mathematics cannot be applied to the entire structure. Is mathematics a limited, man-made tool that is incapable of simultaneously describing these disparate, regional behaviors? Is there a universal mathematical principle that applies at all scales or is the aggregate of the disparate behaviors beyond the reach of mathematics?

Louis deBroglie offered, what has now become, the contemporary interpretation of this dilemma. He proposed that the electron has wave-like properties. The electron does

Energy, Cold Fusion, and Antigravity

not accelerate around the nucleus, but rather, it encircles it in the form of a standing wave. The deBroglie wave is a curious mathematical formulation that shrinks and swells with speed. It has no classical analog. No explanation was provided as to why the undulating deBroglie waves do not leak energy through a process of continuous radiation. The problem of the stability of the atom was, in effect, transferred from the orbiting electron to the undulating deBroglie wave.

A particle-like photon is emitted as the deBroglie wave instantaneously collapses. How do quantum systems instantaneously contract within the bounds of Special Relativity? Does the universe split with each collapse? Max Born attempted to get around these difficulties with his Copenhagen Interpretation. It states that matter's deBroglie wave is not real. Matter's deBroglie's wave is only an construct of probability. It exists, only on paper, within a mathematical configuration space.

Einstein, Podolsky, and Rosen rejected the abstract nature of this construct and stated their objections in the, now famous, E.P.R. paper.[11] Einstein maintained, until his death, that the theory of quantum mechanics was incomplete. The quantum condition was not reconcilable with Newtonian physics.

The renowned Dr. Richard P. Feynman summed it up best in his lecture.

"But how can it be like that? Because you will get 'down the drain', into a blind alley from which nobody has escaped. Nobody knows how it can be like that."

Energy, Cold Fusion, and Antigravity

OUT OF FEYNMAN'S DARK ALLEY

This author introduced his "impedance matched interpretation" of quantum physics. This interpretation, unlike Schrödinger's wave equation, is founded on a classical unifying concept. This concept is that of impedance matching. The quantization of matter and energy classically emerge from the resulting structure. The results are real, not probabilistic, and they do not require a multitude of universes. The interpretation has shown a way out of Feynman's dark alley. This tendency is displayed graphically on page 83.

A ball bouncing off of the Earth is an example of a non-impedance (pronounced em-pee-dance) matched system. The heavy Earth does not rebound from the impact of the light ball. The ball relinquishes its energy, through friction, in a series of progressive bounces. The sodium coolant within a fast breeder reactor is another example of a non-impedance matched system. Neutrons bounce off of the heavy sodium nucleons without imparting much energy. The neutrons are not slowed (thermalized) by the interaction. They maintain enough kinetic energy to initiate the fuel breeding fission reactions.

A billiard ball directly impacting another billiard ball is an example of an impedance matched system. The incoming ball stops and the outgoing ball flies away with all of the kinetic energy. One snap of sound is emitted. The kinetic energy is transferred promptly; without bounce. The speed of the exiting billiard, neglecting friction, is that of the incoming billiard.

Energy, Cold Fusion, and Antigravity

Electrons attempt to take all possible paths into the nucleus. Hugh Everett suggested that the universe splits and every possible path is realized within a multitude of universes. Atomic transitions emit a single photon; not a progressive series of lower energy photons. The emission, of a single photon, demonstrates that quantum transition is a prompt, single step process. A complete transfer of energy, within a single step, is a characteristic of an impedance matched system. The energy of the quantum transition does not flow through a multitude of universes. It promptly flows through a single path of matching impedance. Electrons do not actively bounce. They transit, at least action, within a single step. Electrons transit at points of matching impedance.

In the late 20th Century Frank Znidarsic observed a speed of 1,093,850 meters per second within some low energy nuclear and gravitomagnetic experiments. He concluded that this is the speed of sound within the nucleus S_n. Energy flows from the quantum domain into the classical reality at this speed. The mathematical quantification of this speed shows that the quantum domain emerges as the non-conserved components of the wave function instantaneously collapse and the conserved components interact at a speed of 1,094,000 meters per second.

When an irresistible force meets an unmovable object something has to give. The properties that yield, during the quantum transition, are the non-conserved properties. The non-conserved magnetic components of the force fields are profoundly affected. The electrical magnetic force, the gravitomagnetic force, and the nuclear magnetic forces are ejected from the system. The flux is driven to surface of a

Energy, Cold Fusion, and Antigravity

quantum state where it is forced into a convergence of range and strength. The action resembles that of a magnetic amplifier where a strong magnetic field, which is applied from the outside of its core, effectively destroys the inductance of it's the core. Likewise, a decoupling of the gravitomagnetic flux from matter destroys the matter's inertial mass. Special Relativity demonstrates that inertial mass is not a conserved property. It increases with the addition of speed and it may be mutable under other conditions. Impedance is a function of mass and speed. The inertial mass of matter mutates during a quantum transition, the impedance of the systems is matched, and the quantum jump proceeds promptly at the speed S_n.

Energy, Cold Fusion, and Antigravity

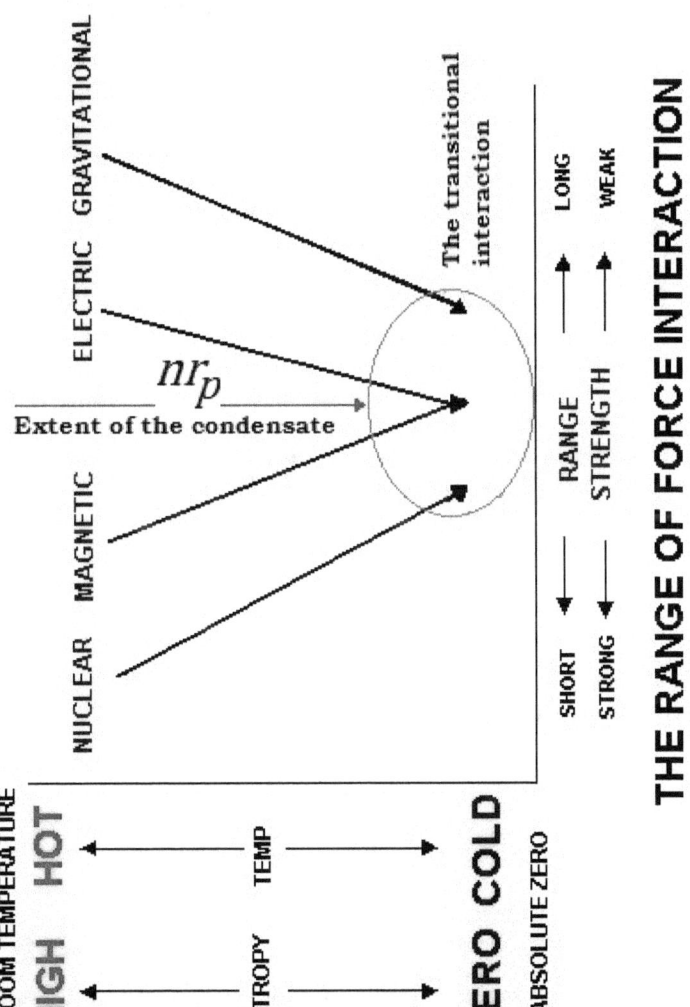

Energy, Cold Fusion, and Antigravity

Chapter 6 The Mathematics

A Refactoring of Coulomb's Equation

In the 18th Century Charles-Augustin de Coulomb gave the force produced by interacting electrical charges (1). This force varies inversely with the square of the displacement **r²**.

(1)

$$Force = -\left(\frac{Q^2}{4\pi e_o}\right)\frac{1}{r^2}$$

The mathematical operation of the integration of (1) gives (2) the energy of interacting electrical charges. This energy varies inversely with the displacement **r**. (2)

$$Energy = \left(\frac{Q^2}{4\pi e_o}\right)\frac{1}{r}$$

James Clerk Maxwell extended Coulomb's result and went on to derive the speed of light **c**. Coulomb's and Maxwell's formulations were a crowning achievement of classical physics. These results describe much of man's contemporary electrical technology.

The electrical field does not appear to be compressed, stretched, or distorted. How does a rigid field exert a

Energy, Cold Fusion, and Antigravity

smoothly acting force? Where does an inelastic field contain its energy? Why are all electrons identical? Is there a supersized electron that standardizes the state of all electrons?[49] Why doesn't the frequency of the electrical field's energy emerge as an intrinsic property of the electric field? How does an inelastic field support quantum oscillations? Contemporary constructs state that the field's force is carried by virtual photons. It is difficult to understand how virtual photons can carry the negative momentum associated with an attractive force. The supposition assumes that quantum realm is the true reality and that the classical realm emerges from the larger quantum reality, like an illusion, as it agglomerates. Today, one Century later, scientists no longer question these difficulties and they invoke Planck's constant ubiquitously.

The parameters of the classical universe are bounded at maximums. For example, there is only so much money, energy, and time in your life. Likewise, the elasticity of a spring is limited by the strength of its material. It breaks down when subjected to a force that is beyond its limit. The force exerted by two electrical charges, compressed to within one classical radius $2r_p$ of each other, is F_m (**29.0535 Newtons**). This author proposes that the rigid electric field buckles when subjected to the force F_m. The theory of a point-like charge being able to buttress an infinite amount of force fails at half the classical radius of the electron.

The electrical force was quantified as the product of a variable pressure ($F_m/\pi r^2$) acting upon the area of a section of a bubble $\pi(2r_p)^2$. The form factors (π) cancel within the solution (3). Equations (1) and (3) yield identical forces;

Energy, Cold Fusion, and Antigravity

however, eq. (3) exposed an action. The electrical field produces a force in reaction to an elastic bubble. This bubble is propelled in a direction that reduces the intensity of the composite field. No negative momentum is required. (3)

$$Force = -\frac{F_m}{r^2}(2r_p)^2$$

The integration of the equation (3) with respect to r gave (4). Equation (4) was expressed in the form of two superimposed mechanical springs ½ **Kx²**. (4)

$$Energy = \tfrac{1}{2}\left(2\frac{F_m}{r}\right)(2r_p)^2$$

Equation (4) gives the energy of two interacting electrical charges. The electrical field buckles when subjected to the force F_m. The buckled field linearly opposes a constant force. The rest energy of the electron is contained within two elastic discontinuities $2r_p$ as a linear product with this force F_m. The electric field buckles sooner at radii **r** where it is weaker. The pliable elastic discontinuities act at a reduced force at these radii. The action fixes the rest energy of a local electron and contains the potential energy of a remote electron. The elastic constant **(F_m/r)** appears as a factor of the motion. The emergence and the polarity of the charge may be due to the inclusion or exclusion of some trapped flux. These effects standardize the state of the electron[47].

Energy, Cold Fusion, and Antigravity

The electron behaves like a ball of plumber's putty, it does not bounce, it promptly surrenders all of its kinetic energy, and it is reconfigured through a process of elastic failure. The quantification of this behavior, in (4), produced the same potential energy as did Coulomb's equation (2). Equations (2) and (4) both describe a large portion of classical reality.

The spring like formulation (4) introduced the elastic factor (F_m/r). Like that of a rubber band, this elastic factor diminishes inversely with the displacement r. The elastic constant, of a spring, expresses a relationship between force and displacement. This restoring force, acting upon a mass, produces a resonant harmonic motion. This motion oscillates in time. Matter's wave-like properties emerge as a classical effect of the elastic factor.

The wave number converts a length into an angle. The angle can then be read by a trigonometric function. Equation (4) introduced the length $2r_p$. This length was used to set the wavelength λ of the electrical field's wave number $(2\pi/\lambda)$. The wave number describes amplitudes that undulate with position. Matter's particle-like properties emerge as a classical effect of the elastic discontinuity. These results, taken together, produced Planck's constant and reconciled the duality of matter and waves.

The domain of equation (4) extends smoothly across the exponential classical reality and then, at the radius r_p, it encounters an elastic discontinuity. The quantum domain appears linearly and smoothly on the inside of the discontinuity. The elastic factor (F_m/r) acts linearly within the linear, quantum domain. Assumptions of linearity are no longer required. The electrical field really is linear at small displacements. The linear relationship, described by the photo-electric effect emerges with this action. The smooth regions can be described with mathematics; however, the discontinuity at their boundary was discovered and imposed. These quantities did not emerge from any anthropic, mathematical analysis. Their existence, outside of the mathematics, shows that the natural realm is beyond mathematics. It also shows that that mathematics was invented, not discovered, and is just a useful tool. The imposition of an elastic boundary replaced many abstract quantum principles, eliminated quantum paradoxes, and precisely defined the limits of the quantum domain. A homogenous description of reality was produced.

The reciprocal of a maximum of mechanical elasticity **K** corresponds with a minimum of electric capacitance **C**. (**C α 1/K**) The elastic discontinuity of a radius r_p appears as an effect of a universal minimum (quantum) of stray electrical capacitance. It takes time to charge a capacitor; however, capacitance is independent of time. This geometric effect is imposed within tiny domains. The quantum domain exists, within a subset of classical reality, as a consequence of this imposition. An extrapolation of equation (4) produced new results. These results include methods to control all of the natural forces and techniques that could reduce inertial mass. The technologies that may come from these understandings could take mankind to the stars.

Energy, Cold Fusion, and Antigravity

The Speed of Hydrogen's Electrons

A solution of the classical wave equation is the function (5). It describes the amplitude **Y** of a traveling wave in space and time. Its first term is called the wave number **($2\pi/\lambda$)**. The wave number describes a wave's amplitude at point **x**. Wave numbers exist, within traveling waves, at integers of the fundamental length **n**. The second term contains the angular frequency **ω**. The angular frequency describes an oscillation at time **t**. Harmonics **n** of the angular frequency do not exist within traveling waves. (5)

$$Y = \cos\left(\frac{2\pi n}{\lambda}x - \omega t\right)$$

A wave propagates along with its wavenumber. The wave number is expressed by the first term in (5). The wave number was set to the length **$2r_p$** in (6). This length is that of the classical radius of the electron. It describes traveling waves. Bosons travel along with this wave number. The angular frequency **ω** equals the square root of **K** over **M**.

(6)

$$Y = \cos\left(\frac{2\pi n}{2r_p}x - \sqrt{\frac{K_{e^-}}{M_{e^-}}}\,t\right)$$

A standard solution of (6) is the angular frequency divided by the wave number. Divide by taking the reciprocal of the wave number times the square root. The result (7) is a product of wavelength and frequency. It gives the speed S. The elastic factor K_{e^-} was expanded in (7) to the ratio of F_m over the ground state radius of the hydrogen atom r_h. The resultant angular frequency (the square root) is the Compton frequency of the electron. (7)

$$S = \frac{2r_p}{2\pi n}\sqrt{\frac{F_m/r_h}{M_{e^-}}}$$

Equation (7) was simplified and the right side was multiplied by 2π to give the circumferential speed in (8). There is no component of this speed propagating the radial direction. The state is known, because of this lack of radial motion, as a stationary quantum state. (8)

$$S_{e^-} = \frac{2r_p}{n}\sqrt{\frac{F_m/r_h}{M_{e^-}}}$$

The result S_{e^-} equals 2.19 million meters per second divided by the number n ($n = 1, 2, 3...$) of the atomic orbit. These speeds are those of the orbiting electron in the hydrogen atom. The result demonstrates the efficacy of the elastic factor and the wave number. The orbiting electron induces an electromagnetic magnetic moment. No strong gravitational or long range nuclear forces emanate from an orbiting electron. The quantized orbital spin of the electron emerged as a function of the classical wave equation.

Energy, Cold Fusion, and Antigravity

The Speed of Sound in the Nucleus

Niels Bohr described the nucleus, in 1936, as a stiff, high density (2.3×10^{17} kg/m³) drop of liquid. The field of nuclear physics was developed greatly from that time, however, the speed of sound in the nucleus was never seriously considered. As with all substances, the nuclear fluid conveys sound at a speed based upon its density and elasticity. The density of the nucleus is known to be constant. The nuclear density is independent of the number of neutrons or the number of protons. This implies that the forces that act upon the nucleons are nearly constant. This consistency in density and elasticity fixes the speed of sound within the nucleus.

The function (9) describes a standing wave as the product of the cosine of its wave number and the sin of its angular frequency ω. Notice that a traveling wave was formulated as a difference in (5) and that a standing wave was formulated as a product in (9). These functions emerge as solutions of the classical wave equation and are a standard. Harmonics **n**, of the natural frequency ω of a standing wave exist. Harmonics of the wave number do not exist. (9)

$$Y = \cos\left(\frac{2\pi}{\lambda} x\right) \sin(n\omega_n t)$$

The strong nuclear force acts upon itself. This action can be either attractive or repulsive. The stability of the nucleus is maintained by a balance of these opposing actions. The elasticity of the nuclear force, at these balance points, is zero. High energy physicists find these balance points to be

Energy, Cold Fusion, and Antigravity

disinteresting because the constituent quarks are relaxed and undetectable. This author; however, has discovered that this quiescence betrays the stability of matter. The nucleus also feels the electromagnetic force. A quantum of stray electrical capacitance establishes a floor in the elasticity of the nucleus. The residual elasticity of the nucleus is the sum of its electrical elasticity and the elasticity of a counteracting nuclear force. The angular frequency of the nucleons ω_n was determined by taking the square root of twice the electrical elastic factor K_e- divided by the proton's mass M_p. This factor was inserted into (9) producing (10).

The nucleus is built of alpha particles as a wall is made of bricks. The nucleons, within an alpha particle are crushed to a point where the strong nuclear force becomes repulsive. The nucleon spacing has been determined by the scattering of these alpha particles.[34, 37] This spacing K_f equals 1.36 Fermis. The reciprocal of this spacing is known as the Fermi wave number. The length of the nuclear spacing K_f was inserted into the wave number in (9) producing (10).

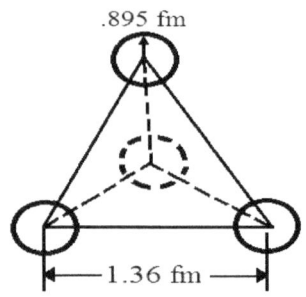

The tetrahedral arrangement of bound nucleons

Energy, Cold Fusion, and Antigravity

(10)

$$Y = \cos\left(\frac{2\pi}{K_f}x\right) \sin\left(n\sqrt{\frac{2K_{e^-}}{M_p}}\,t\right)$$

A solution of (10) is again the angular frequency divided by the wave number. It gave a speed in (11). The elastic factor K_{e^-} was expanded and set to the Fermi displacement length K_f. The longitudinal motion of a sound wave does not require an additional factor of 2π. Equation (11), expresses speed as the product of a frequency $[(1/2\pi)\omega]$ and the displacement K_f. The result is the apparent speed of sound in the nucleus. (11)

$$1{,}093{,}850 \; m/s = \frac{n}{2\pi}\sqrt{\frac{2F_m/K_f}{M_p}}\,K_f$$

The speed of a wave varies with its frequency in some media. A prism employs this type of motion. The wavelength of a nuclear disturbance is fixed at the displacement K_f. Speeds appear with the harmonics **n**. These harmonics are associated with the atomic number **Z** of the nucleus.

The speed of sound in the nucleus was determined between protons at the Fermi scale. The consistency of the nuclear density and elasticity implies that this speed S_n is independent of scale. This speed is also exhibited within an isolated nucleon and at the atomic scale within the active environment of a low energy nuclear reaction.

Energy, Cold Fusion, and Antigravity

It is also appears within the electron as the product $2\pi f_c r_p$. More generally, the speed is that of a longitudinal matter wave.

The transitional speed appears to be only slightly affected by the presence of neutrons. The velocity is maintained as the weak force independently resonates with the transitional velocity. The strength of the atomic chemical bonds increases slightly with the number of neutrons. This relationship is known as the kinetic isotope effect. An increase in the strength of the atomic bonds is the reason why heavy water is toxic. The relationship suggests that the quantization of the atomic orbits is established through an action of the nucleus.

This author's impedance matched interpretation of quantum physics quantifies this relationship. It states that $2\pi n$ times the speed of a transitional electron equals a harmonic Z of the nuclear speed S_n. The factors (n and Z) are integers. Integers appear within the descriptions of classical harmonic motion.

$$ZS_n = 2\pi n \text{ (speed of the transitional electron)}$$

The longitudinal speed of a matter wave is a fundamental classical constant. The quantum condition emerges as an effect of its action.

Energy, Cold Fusion, and Antigravity

THE RADAII OF THE HYDROGEN ATOM

Maxwell's theories predicted that accelerating electrons will continuously emit electromagnetic radiation. This loss of energy will drive a spiraling electron into the nucleus. Bound electrons do experience a constant centripetal acceleration; however, they do not continuously emit energy. Atoms emit random bursts of energy as their electrons jump between a set of allowed atomic orbits. These orbits were quantified, by Bohr and Schrödinger, at radii containing a reduced quantity (**h/2π**) of Planck's angular momentum. This qualification has no classical analog and, as such, has led us down into Feynman's dark alley. There we have remained stuck for over 100 years.

An oscillation at the angular frequency of ω is produced as an elastic medium tugs upon a mass. The frequency of this type of motion is routinely quantified as the square root of the ratio of an elastic factor (**K$_{e-}$**) to a mass (**M$_{e-}$**). The fundamental displacement of a standing wave is one-half wavelength. The length is expressed by the wave number in (12). The length **r$_p$** is that of the non-rigid elastic discontinuity. Fermions oscillate within this elastic zone.

(12)

$$Y = \cos\left(\frac{2\pi}{r_p}x\right) \sin\left(n\sqrt{\frac{K_{e-}}{M_{e-}}}t\right)$$

Energy, Cold Fusion, and Antigravity

Standing waves, by definition, do not propagate. A solution of (12) (again the angular frequency divided by the wave number) gives a speed in (13). The electron moves between the harmonics **n** of the natural frequency at the atomic speed S_{e^-}. The principle quantum number **n** emerges as an effect of these harmonics.

(13)

$$S_{e^-} = \frac{n}{2\pi} \sqrt{\frac{K_{e^-}}{M_{e^-}}}\, r_p$$

If you are bored or lost, hang on, we are getting best part of this story. To understand the importance of what we are about to do, you have to go back to early 20th Century. In 1913, as Niels Bohr was presenting has quantum atomic model Ernest Rutherford wrote,

"There appears to me one grave difficulty in your hypothesis, which I have no doubt you fully realize, namely how does an electron decide what frequency it's going to vibrate at when it passes from one stationary state to another? It seems to me that you have to assume that the electron knows beforehand where it is going to stop."[32]

In 20th and 21st Century a great effort was made building accelerators in an attempt to resolve Rutherford's quandary. No solution was discovered within the resulting high energy reactions. Surprisingly, a solution was gleamed from the observation of low energy nuclear reactions. This solution emerged at the intersection of functions (11) and (13).

Energy, Cold Fusion, and Antigravity

The speed of a transverse electrical wave S_{e^-} was set equal to the speed of the longitudinal matter wave S_n. The speed of the transverse electrical wave is that of a jumping electron. In more general terms the speed of light, in the atom, was set equal to the speed of sound. The result (14) determines the frequency of the transitional electron, fixes the energy levels of the atom, shows that the ground state orbit cannot radiate, and matches the impedance of the interacting partners. It exposes the most fundamental action of the natural realm. How was this missed for so many years?

The elastic factor in (13) was expanded in (14). A factor of 2π was injected into the right side of (14) as per this author's impedance matched interpretation of quantum physics. Hydrogen's electron interacts with the fundamental nuclear speed. (14)

$$S_n = n \sqrt{\frac{F_m/r_a}{M_{e^-}}} \, r_p$$

The elastic discontinuity r_p attaches a superposition of electronic states to an atomic energy level. Transitional vibrations stimulate the single attached state until the elastic limit of the bond is exceeded. The electron decoheres and flows through a channel of matching impedance. A second electronic energy level then coheres from another superposition of states. The channel of matching impedance is characterized by its speed. There is no component of this speed propagating circumferentially around the atom.

Energy, Cold Fusion, and Antigravity

The state moves radially inward and radially outward with a change in the harmonic **n** of the natural frequency. The speed of this radial movement is the speed S_n. The radial motion is one of quantum transition. Equation (14) described the speed of the radial motion.

Bohr's construct, of the quantization of angular momentum, harmoniously emerges with the action expressed in (14). Equation (14) was solved for the unknown atomic radii r_a in (15). (15)

$$r_a = n^2 \left[\frac{F_m \, r_p^2}{S_n^2 \, M_{e^-}} \right]$$

The length .529 angstroms emerged as the terms within the brackets [ ] were reduced. This length is that of the ground state radius r_h of the hydrogen atom. Hydrogen's principle orbits r_a exist at integer squares n^2 of the ground state radius. The radii of the hydrogen atom depend upon the speed of sound in the nucleus. (16)

$$r_a = n^2 r_h$$

The electron attempts to take all paths into the nucleus. The only open paths are ones of matching impedance. These paths were quantified in terms of their speed. This speed is that of the transitional quantum state. The principle energy levels of hydrogen emerged, from speed, within a subset of Newtonian mechanics.

Energy, Cold Fusion, and Antigravity

THE RADAII OF THE HEAVY ELEMENTS

The effect of multiple overlapping electric charges **Z** increases the elastic factor K_{e^-} of the system by a factor of **Z**. This factor of **Z** was inserted into (12) and produced (17). The wave number, as expressed by r_p, does not vary with atomic number or frequency. It is an independent natural constant. The harmonic motion associated with heaver, high **Z**, elements was described in (17).

(17)

$$Y = \cos\left(\frac{2\pi}{r_p} x\right) \sin\left(n \sqrt{\frac{ZK_{e^-}}{M_{e^-}}} t \right)$$

A solution of (17) (again the angular frequency divided by the wave number) gave a speed in (18). A factor of **2π** was injected into the right side of (18) as per this author's impedance matched interpretation of quantum physics. **Z** is selected by an orbital restriction; where the electron's frequency couples with a harmonic of the transitional speed.

(18)

$$ZS_n = n \sqrt{\frac{Z(F_m/r_a)}{M_{e^-}}} \; r_p$$

A solution of (18) produced (19) the atomic radii r_a of the single electron higher atomic number **Z** atoms.

Energy, Cold Fusion, and Antigravity

The principle energy levels of every element harmoniously emerged with the square root of a variable elastic factor.

(19)

$$r_a = \frac{n^2 r_h}{Z}$$

The flash and the bang of the quantum transition propagate simultaneously. A single photon is emitted in concert with this coincident action. The motion was quantified by an equality in speed.

The energy of the natural forces participates in the prompt, bounce-less transition. The strength and the range of the magnetic components of the natural forces are driven into a convergence. The kinetic energy of the transitional electron is balanced by the negative potential of an induced strong, local gravitomagnetic field. The channel of matching impedance extends across the length of a quantum system. The nuclear forces participate in the reaction. Nothing rattles.

The diamagnetic condition reduces the inertial mass of the transitional electron. The mass energy flows between the energy levels of the atom at 1,093,850 meters per second (S_n). The principle atomic energy levels emerged, from speed, within a subset of Newtonian mechanics.

Energy, Cold Fusion, and Antigravity

QUANTUM SPIN

In 1922, German physicists Otto Stern and Walther Gerlich conducted an experiment that measured an electron's spin. Electrons were passed through a powerful magnetic field. A spinning electron induces a tiny magnetic dipole. They expected that the electrons would enter into the external magnetic field with their magnetic dipoles pointing at random angles. The interaction of the applied field, with the electron's magnetic dipole, would then deflect the electrons at random angles. They discovered something that was quite unexpected; the electrons were all deflected either up or down at the same angle. The electron appeared to be spinning either up or down. No electrons, spinning at other angles, were observed. They then preset the spins with another magnet prior to sending them to the detector. As before all of the electrons emerged from the second detector either in the spin up or in the spin down direction. The preset condition, of the electrons, only affected the probability of detecting a spin up or a spin down direction. The result seems to be influenced by the process of observation. No existing classical model could explain this one dimensional, binary, and uncertain behavior. It was said that the behavior demonstrated that the quantum domain existed outside of the domain of Newtonian mechanics. It has been suggested that the universe splits with the emergence of one of these outcomes. The behavior was placed within an abstract realm of mathematical Hilbert Space.

Energy, Cold Fusion, and Antigravity

The impedance matched interpretation describes quantum interactions in terms of their speed. These speeds, more correctly, velocities, act orthogonally (at right angles ⊥) to each other. Equation (18) quantified the magnitude of the electron's intrinsic spin. Vector notation was inserted into (18) and the cross product (20) was produced. Equation (20) describes both the magnitude and the direction of the electron's spin. S_n points in the direction of quantum transition. The unit vector (⊥-hat) points radially from the nucleus. S_n and (⊥-hat) may or may not be orthogonal. The vector r_p points in the direction of the electron's velocity. The vector r_p is always orthogonal to (⊥-hat). The cross product (⊥-hat) x r_p points in the direction of the electron's spin. Equation (20) shows the orthogonal orientation of the motions. (20)

$$Z\vec{S_n} = n_1 \sqrt{\frac{ZF_m/(n_2^2 r_g)}{M_{e^-}}} \hat{\perp} \times \left(\pm \vec{r_p}\right)$$

The product of the integer squared n_2^2 and the ground state radius r_g sets the electron's elastic factor. This product expresses the atomic radii. Classical, integer harmonics n_1 appear with the square root these radii. The squared atomic radii emerge harmonically as an effect of the variable elastic factor.

The external magnetic field within the Stern Gerlich experiment prepares the system by orthogonally aligning the vectors S_n and r_p.

Energy, Cold Fusion, and Antigravity

Contact with matter perturbs the aligned system. S_n selects a direction and begins to travel. This motion establishes the direction of the electron's spin.

This author's model begins with the classical wave equation. The classical wave equation is a second order differential equation. The existence and uniqueness theorem states that the solutions of second order differential equations come in pairs. In the case of the classical wave equation these elements are of opposite sign. The process of observation can set up the boundary condition that initiates a cascade. The direction (+ or -) of S_n is selected. The second solution of classical wave equation is suppressed. It remains within the quantum realm and preserves the angular momentum of the broken superposition. No split in the universe is required. God appears to roll the dice with this selection. The one dimensional, binary, and uncertain behavior of a quantum system appears naturally with this multiplicity. This uncertainty has been contemporarily been expressed in units of angular momentum as $(\pm h/4\pi)$. The Stern Gerlich appears as an observable effect of this uncertainty. Quantum uncertainty as defined by the Heisenberg Uncertainty Relationship and the quantum measurement paradox emerge as global effects of this duality. Time flows within the bounds of this uncertainty. The probabilistic nature of the quantum realm appears at the instant a standing superposition of a solution to the classical wave equation is broken. The wave function then collapses leaving in its wake a host of real observables. The intrinsic spin of the electron harmoniously emerged as a function of the classical wave equation.

Energy, Cold Fusion, and Antigravity

This author introduced an elastic factor and a wave number. The electron's orbital spin (5) steps by the way of two elastic discontinuities $2r_p$. This orbital state is not elastically deformed and it is not pinned at a central discontinuity. It freely flows around the electron. The energy of these non-bound structures can bounce. They tend to bounce away their energy and to congregate at the lowest energy levels. These structures display a spin of one and they obey Bose Einstein Statistics. The spin one-state appears as an effect of a wave number with a length equal an even multiple of r_p.

The intrinsic spin of the election emerged in (12) with this author's wave number r_p. The electron elastically deforms when compressed to the radius r_p. The electronic states are pined at the emergent discontinuity. The ensemble destructively interferes with its parts. No two parts can occupy the same state. These structures display a spin of one half and they obey Fermi Dirac Statistics. The rigidity of matter as described the Pauli Exclusion Principle is fixed by this action. The spin-1/2 state appears as an effect of a wave number with a length equal an odd multiple of r_p. These foundations are not as simple as those of Bohr model because they do require a factor of **Z**; however, they show that quantum discreteness emerges with a classical set of wave numbers and an elastic constant.

The vector r_p may be orthogonal to S_n. An equality in the factors (20) of **n** appears with the real orthogonal orientation of all of the vectors. The electron vibrates at its Compton frequency under this condition. This is the state of the spherical **S** atomic orbit. The S orbital contains two orthogonal states. One is spin up ($+r_p$) and the other is spin

Energy, Cold Fusion, and Antigravity

down ($-r_p$). The principle atomic orbits ($n_1, n_2, n_3...$) are described by equalities in the factor of **n**.

The orientation of the vectors S_n and r_p may not be orthogonal within a bound state. The fine atomic structure emerges with this freedom. A harmonic of the Compton frequency n_1 can be one within the second principle ($n_2=2$) atomic orbit. The Compton frequency of the electron downshifts under this condition. The tilting in the axis of electron's orbit maintains synchronization with the invariant transitional velocity. Peanut shaped orbits appear with this tilt. Schrödinger's wave equation does not address the downshifting of the Compton frequency of the electron. It will be shown that this downshifting is of universal importance. It is responsible for the fine atomic structure, low energy nuclear reactions, and gravitational time dilation. The precession (wobbling motion), of the orbit, is fettered to a fraction of the transitional velocity. The dynamics resembles that of two wobbling tops each on opposite sides of a plane. The tilt of the orbit is traced by the precession of the motion. The tilt of the orbit is described in (20) by the angular momentum quantum number *l*, where ($l = n_2 - n_1$). Two **P** orbits of opposite spin fit into three orthogonal positions. The sum of the two **S** states and the six **P** states equals eight points of matching impedance. The second principle orbit ($n_2=2$) of the atom can hold eight electrons. When filled, the eight electron structure is chemically stable. The analysis can be extended to all of the atomic orbits. The fine atomic structure emerges with the orbital precession. The fine structure and its uncertainty harmoniously emerged within a downshifted version of the classical wave equation.

THE SPECTRAL INTENSITY

The spectral lines, emitted by glowing matter, vary greatly in intensity. Bohr's semi-classical atomic model could not account this variation in intensity. Werner Heisenberg offered a solution that arranged quantum properties on a matrix. Planck's empirical constant was inserted abstractly, by Heisenberg, into the formulation as a commutative property of matrix multiplication. Heisenberg's application of vector mechanics to quantum physics is considered to be one of mankind's greatest intellectual accomplishments. Heisenberg's solution gave the intensity of the spectral emissions and established the field of quantum physics. The formability did not, however, expose the underlying action.

Louis deBroglie proposed that matter is a wave[42]. Erwin Schrödinger incorporated deBroglie's wave into his wave equation. Schrödinger's result also produced the intensity of the spectral emissions. Schrödinger's introduction, of the deBroglie wave, produced a clearer solution but, in the process, it introduced many conceptual problems. Schrödinger's wave equation is scalar. Scalar equations do not incorporate any restraining forces. What forceless mechanism bundles the deBroglie wave into to a particle? How does a wave couple with this point-like bondage? Bohr's principle of quantum correspondence was invoked in an attempt to justify why the energy of a classical wave is associated with its amplitude and the energy of a quantum wave is associated with its frequency. The entire construct revolves around the abstract, empirical Planck's constant. Heisenberg and Schrödinger knew nothing about the path of

Energy, Cold Fusion, and Antigravity

the quantum transition. Their solutions did not directly incorporate a probability of transition.

Max Born's Copenhagen Interpretation suggested that the mater wave is restrained by a Fourier superposition of an infinite number of probability waves. The probability of a particle being in a particular location is proportionate to the amplitude of these subjective waves of probability squared.

"By which strange coincidence could a representation of the probabilities propagate in space through time like a physical wave able to be reflected, refracted and diffracted?"

Louis deBrogle

Max Born did not extend the Copenhagen Interpretation beyond the electromagnetic force and; for this reason; a classical explanation of quantum transition did not emerge from the model. This author's Impedance Matched Interpretation does extend to all of the natural forces and it commences with the classical wave equation.

In 1916 Einstein published the famous paper "Emission and Absorption of Radiation in Quantum Theory". This paper stated that the probability of transition could be increased through the action of an external stimulation. This stimulative process was applied, in a limited way, to the development of the LASER. Znidarsic observed a stimulative process at work within low energy nuclear reactions. The action of this stimulative process is universal. The energy of the matter wave transitions at the sonic nuclear speed S_n (1,093,850 meters per second). The probability of quantum transition, not a particle's location, is proportionate to the square of the amplitude of vibration at the speed S_n. The gravitomagnetic, electro-magnetic, and nuclear magnetic force fields are driven out of the system and the range and the strength of the force fields converge. This vibration destroys inertial mass and opens a channel of matching impedance. Atomic transitions progress through these open channels. The transitional action is enhanced within the boundaries of the double slit experiment. The motion, with more transitional vibration, is that of a disperse wave. This wave pilots the electrons pass through the slits.

In review, this author has shown that the stationary atomic radii r_a exist at points of matching impedance (15). The transitional atomic surface waves are compressed between these surfaces. This compression increases the speed of these surface waves to S_n, (15)

$$r_a = n^2 \left[\frac{F_m \, r_p^2}{S_n^2 \, M_{e-}} \right]$$

The impedance matched interpretation of quantum physics describes the velocity of quantum transition (22). This multi-force-field action opens a channel of matching impedance and emits a photon. Chemically assisted nuclear reactions are induced by an intense vibration at this dimensional frequency.

(22)

$S_n = 2\pi n$ (light speed in an atom)

Energy, Cold Fusion, and Antigravity

The speed of light in an atom was refactored (23) as the product of the atomic frequency f_a and the circumferential length of a transitional state $2\pi r_a$. Harmonics **n** of the frequency f_a exist. (23)

$$S_n = (2\pi n f_a) r_a$$

The union of equations (15) and (23) is (24). It describes the amplitude r_a of the transitional vibration in terms of the observable frequency f_a of the emitted light. The frequency of the light emitted by an atom is not generated by the vibration of the stationary atomic orbits. It is, however; that of the transitional vibration. The frequency of this vibration f_a exhibits duality in that it energetically couples the stationary atomic state **n** to another state. (24)

$$r_a = n^2 \left[\frac{F_m r_p^2}{(2\pi n f_a) r_a S_n M_{e^-}} \right]$$

The amplitude squared r_a^2 appears in (25) with the reduction of (24). This amplitude is that of a transitional quantum state. The series of a group of spectral emissions is represented by the integer **n** (ref. plot on page 111).

(25)

$$r_a^2 = n \left[\frac{F_m r_p^2}{(2\pi f_a) S_n M_{e^-}} \right]$$

Energy, Cold Fusion, and Antigravity

The constants in equation (25) were regrouped in (26) and the numerator and denominator were multiplied by a factor of **4π**. (26)

$$r_a^2 = \left[\frac{4\pi F_m r_p^2}{S_n}\right] \frac{n}{8\pi^2 M_{e^-} f_a}$$

The reduction of the factors within the brackets [ ] produced Planck's constant. Planck's constant emerged as a dependent variable. The result (27) is the well-known formulation for the amplitude of harmonic motion. The electronic states are associated or disassociated with an elastic discontinuity through the action of the transitional vibration. This action was quantified in (27) as the algebraic union of two real equations. The transitional action applies both to the quantum jumps between an atom's inner states and to the classical motion of the atom's outer electrons. No principles of correspondence or quantum fudge factors were required. The probability of transition varies directly with the square of the real amplitude, r_a^2. The intensity of a spectral emission varies with this probability. Perhaps (27) should be solved for the dependent variable f_a and renamed, "The Frequency of the Transitional Amplitudes".

(27)

$$r_a^2 = \frac{nh}{8\pi^2 M_{e^-} f_a}$$

Energy, Cold Fusion, and Antigravity

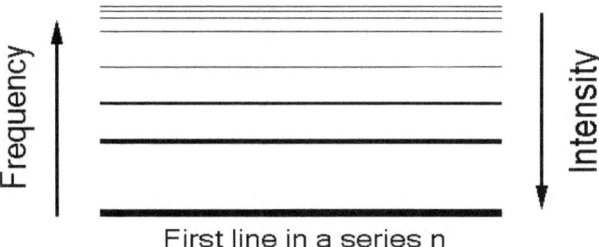

First line in a series n

The intensity of a group of spectral lines

The frequency of a classical wave is that of its emitter. Equation (27) shows that the frequency of an emitted photon is that of the transitional atomic vibration f_a. The intensity of a classical wave varies with the square of its amplitude. The number (not the energy or the voltage) of emitted photons varies with the square of the transitional amplitude r_a^2. The simultaneous emergence of both the photon's frequency and the matter wave's amplitude is fundamental to and replaces Bohr's principle of quantum correspondence.

The strength of the expelled electromagnetic, nuclear magnetic, and gravitomagnetic forces increases with the square of the amplitude of the vibration r_a^2. Inertial mass is reduced as the gravitomagnetic flux is decoupled from the core of matter. The probability of transition increases as the velocities converge and impedance of the fields is matched. The intensity of the light emitted by a bound electron is proportionate to the probability of transition. Time is metered through the action of these probabilities. A real probability of quantum transition emerged from the classical wave equation with the injection of an elastic constant and a wave number.

Energy, Cold Fusion, and Antigravity

Theoretical Cold Fusion

The current paradigm in physics has been to probe at increasingly higher energies. This author's model is collective not reductionist. It adds layers and steps back into lower energies in an effort to discover new principles. It has revealed reactions that are not those of bumping particles but rather which are akin to those of an electrical transformer.

A low energy nuclear reactor (LENR) is constructed of palladium or nickel. Hydrogen or deuterium gas is driven into the metal. The dissolved gas is not bound. It is; however, confined within the bounds of the nanoparticle. This length of the confinement determines the wave number of the dissolved gas. A metallic lattice naturally resonates at thermal frequencies. This vibration can be enhanced with heat or with infrared stimulation. Nanometer scaled domains and cracks strongly couple with the lattice vibrations[35]. This stimulation acting across the bounds of a nanoparticle imposes a speed.

Typically, vibration destroys superconductivity. This author's analysis has shown that vibration, at a dimensional frequency of one megahertz-meter, enhances superconductivity[36]. A similar effect was experimentally discovered at the max Planck institute in 2014[41].

"Terahertz-frequency (10^{12} hz) optical pulses can ... induce superconductivity", Roman Mankowsky, The Max Planck Institute[41]

Vibration can reinforce an electronic condensate and induce a warm, above room temperature, superconductivity. An intense vibration can also adjoin dissolved nucleons to the condensate. The resulting condensate is bound by excitons not cooper pairs.[44] This composite state is reminiscent of J.J. Thompson's "plumb pudding" model of matter.

Low energy thermal vibration has a high energy effect at the scale of a condensed nano-domain. The stimulation can impose a speed of 1,093,850 meters per second within a nanoparticle. The action at this velocity drives the electromagnetic, gravitomagnetic, and the nuclear magnetic forces to the surface of the nano-domain. The system enters into a macroscopic state of quantum transition. *"The constants of the motion tend toward those of the electromagnetic"* and the LENR reaction proceeds.

A notation was used to describe the motion of a standing wave in (9). The quantification of this motion has produced several well-known results. Equation (9) will now be consistently applied and it will show new results. (9)

$$Y = \cos\left(\frac{2\pi}{\lambda}x\right)\sin(n\omega_e \cdot t)$$

The isolated electron resonates at the highest possible frequency; its Compton frequency f_c. The angular frequency ω_{e^-} was set to the Compton frequency of the electron ($2\pi f_c$). The Compton frequency of a condensate, like that of a chime, varies inversely its length. Harmonics appear at fractions (**1/n**) of the electronic Compton frequency.

The length of electronic wave number $\mathbf{r_p}$ emerged from the refactoring of Coulomb's equation. The dissolved gas is confined at a multiple of this length ($\mathbf{nr_p}$). The reciprocal of

Energy, Cold Fusion, and Antigravity

this length was used to set the wave number of the dissolved gas. These conditions were inserted into (9) resulting in (29).

$$Y = \cos\left(\frac{2\pi}{nr_p}x\right)\sin\left(\frac{2\pi f_c}{n}t\right) \tag{29}$$

A solution of (29) (again the angular frequency divided by the wave number) gave a speed in (30). Equation (30) is an identity as both sides of the equation contain independent constants. A factor of **2π** was injected into the right side of the solution as the longitudinal motion of the dissolved gas follows the transverse motion of the electronic cloud. The result (30) expresses the transitional speed S_n.

The normal speed of sound in a metal is about 5000 m/s. This speed appears as each nucleon interacts with its neighbor. The interatomic spacing is about 2 angstroms. The reciprocal of the interatomic length expresses the wave number of a classical solid. When set to the atomic spacing and the nuclear mass (7) yields the speed of sound in a solid. This additional result substantiates this author's model at the slightly larger, classical molecular scale. No adjustments to the motion constants are required at the boundary of the quantum and the classical scales.

Dr. Edmund Storms and Dr. Michael Mckurby have found that a Nuclear Active Environment appears as dissolved protons are driven into a metallic lattice. The resulting pressure increases the frequency of the lattice vibrations. The higher frequency vibrations respond to hotter

Energy, Cold Fusion, and Antigravity

temperatures. The amplitude of the vibration also increases with temperature. The frequency of the intense thermal vibration reinforces the downshifted Compton frequency of the condensate. This reinforcement appears to be adequate to induce a Low Energy Nuclear Reaction. The compression of this motion between the thin layers of Patterson's nano-material drives the surface speed S_n. This speed of these compressed surface waves, like that of a tidal wave, greatly exceeds the speed of sound in a solid. The Low Level Nuclear Reaction speed is difficult to impose; however, it appears to be more accessible than the hot fusion temperature[46]. Generally, this difficulty accounts for the transient stability of matter. (30)

$$S_n = \left(\frac{2\pi f_c}{n}\right)(nr_p)$$

The concept that a superconductor downshifts the frequencies of matter was proposed by Dr. David Noever at the Marshall Spaceflight Center.

"In the models under discussion, the energy gets transferred to vibrations through massive multi-quantum down-conversion."   Peter Hagelstein MIT [48]

This author applied this idea to the Compton frequency of a condensate. The factor, on the left side of (30), expresses the downshifted Compton frequency of the condensate. With **n** set to one there is no downshift and the frequency is the Compton frequency of the electron f_c (10^{20} hz). With **n** set to ten to sixth power the downshifted frequency is in terahertz

(10^{13} hz). This is the frequency, of the thermal vibration, that stimulates a low energy nuclear reaction. With **n** set to ten to the ninth power the downshifted frequency is in megahertz (10^6 hz). This is the frequency, of the radio waves, used to excite Podkletnov's gravitational disk.

The weak force acts over a very short range. It is virtually non-existent at the radius of a nucleon. This author's model suggests that the magnetic component of the weak force is expelled beyond the radius of the neutron. This magnetic component resonates with the elastic discontinuity r_p at the velocity S_n. This motion maintains the transitional velocity within the nucleus. Quantum spin is an effect of the orthogonal action of the expelled magnetic field. The vibration is sufficient to activate Beta decay in an isolated neutron in about 15 minutes. The range of the interaction is expressed by the right factor in (30) with **n** set to one.

Beta decay proceeds within nuclei that contain many neutrons. These neutrons raise the energy of the nucleus and increase the frequency of the transitional vibration. The resulting increase in the transitional velocity within the nucleus is fundamental to the atomic kinetic isotope effect. Nuclear isospin appears as an artifact of this increase in velocity. The probability of a weak nuclear decay increases with the magnitude of this isospin. The inertial mass of the atomic neutrons decreases as the range of the magnetic component of weak force increases. A heavy **W** particle can now be expelled, without borrowing energy, to beyond the surface of a nucleus to the new radius of matching impedance **nr_p**. This action is expressed in (30) with **n** set to greater than one and less than about 150.

Energy, Cold Fusion, and Antigravity

External vibration can also resonate with a cluster of deuterons dissolved with in a crystal lattice. The magnetic component of the weak nuclear force and its associated **W** particle are now expelled to beyond the surface of the cluster. This action is expressed in (30) with **n** set to about one million. LENR Beta decay acts at the range nr_p exceeds the range of the coulombic force and the reaction proceeds smoothly over the nucleus's electrostatic barrier. The vibration imposed by the velocity S_n at this extent is thermal. These cells produce small amounts of highly detectable tritium. The dynamic action may present a high cross-section to neutrinos. The captured neutrinos open a route for the production of tritium. The weak nuclear force may be mutable at very low energies.

Alpha decay is described by equation (30) with **n** set to about two. The energy per emitted per particle at the transitional frequency of the cluster (½ $2\pi f_c$) is about 1.6 MeV. Given that four nucleons are emitted per alpha decay, the total energy of an alpha decay is about 6.4 MeV. The energy obtained by this method is consistent with the accepted values. This consistency of the disintegration energies indicates that they emerge as an effect of single fundamental action. LENR cells that exhibit alpha decay are loaded with hydrogen. The light weight hydrogen vibrates intensely at the range nr_p and induces smooth Alpha transmutations in the heavy nucleons of the crystal lattice. These cells tend to produce helium-4. LENR alpha decay is mediated by the dynamic, magnetic component of the strong nuclear force.

Downshifting to the RF band has been attempted. The resulting RF energy could be directly captured with a diode.

Energy, Cold Fusion, and Antigravity

CHAPTER 7 Mysteries Resolved

THE Photo Electric Effect

Max Planck introduced the quantum and, with it, produced the emission spectrum of a hot blackbody. As a conventional physicist Planck struggled to find a classical explanation for these spectral emissions. Over one hundred years later, the emergence of new observables, has exposed one. The quantization of energy emerges as an effect of a classical impedance match. Energy flows, without bounce, and one photon is emitted. This author's impedance matched interpretation of quantum physics was defined by an equality in speed (31). It states that the speed of a longitudinal matter wave S_n equals 2π times the speed of a transverse electrical wave. The frequency of a traveling wave has no harmonics **n**. This speed is that of a quantum jump.

(31)

$$S_n = 2\pi f_a \lambda_a$$

The frequency of the emitted photon does not match the frequency of any stationary atomic state. As expressed in his letter, Rutherford was puzzled by this. The photon's frequency, does however, match the frequency of the transitional atomic state f_a. The energy of a photon emerges as the radius λ_a couples with an electrical charge.

Energy, Cold Fusion, and Antigravity

The simultaneous emergence of both the photon's frequency and its energy is fundamental to Bohr's principle of complementarity. In combination, these affects reconcile the wave-like and particle-like duality of light and they resolve the measurement paradox.

Albert Einstein, Paul Langevin, Ernest Rutherford, and Max Planck, among others, searched for a classical explanation for the photo electric effect. After a century of looking, none was ever found. In the late 20th Century Znidarsic discovered the speed of sound within the matter wave (11 &31). Matter interacts with light at a speed of 1,093,850 meters/second.

The isotropic capacitance C of a sphere, of radius λ_a was given in (32). Capacitance is a static geometric function. A bubble of capacitance forms within the quantum domain. The proximate components of light immediately collapse upon this geometry during transition. The charge radius decreases as the energy accumulates. Vibration, at the radius of matching impedance λ_a, decoheres particles. Particles interact with waves while in their decoherent state. The particle then transits to another coherent state at the transitional speed S_n. The radius of matching impedance is unknown; however, the light emitted at this radius is observable. The radius λ_a, like the length of an antenna (ref 31), couples the frequency f_a to the speed S_n

(32)

$$C = 4\pi e_o \lambda_a$$

Energy, Cold Fusion, and Antigravity

The solution (33) of equations (31) and (32) expresses capacitance in terms of the frequency of the emitted light f_a and the transitional velocity S_n.

(33)
$$C = \frac{2e_o S_n}{f_a}$$

The energy contained by an electrical charge was expressed in terms of its capacitance in (34).

(34)
$$Energy = \frac{Q^2}{2C}$$

Equation (34) shows that energy varies inversely with capacitance. The voltage produced by an electrical charge increases as its capacitance decreases. The energy of the photon is proportional to the amplitude of this voltage (Power α ExH). This relationship is fundamental to and replaces Bohr's principle of complementarity.

The simultaneous solution of (33) & (34) produced (35). Equation (35) gives the energy of a light wave as a function of its frequency. (35)

$$Energy = \left[\frac{Q^2}{4e_o S_n}\right] f_a$$

Energy, Cold Fusion, and Antigravity

Planck's constant emerged as the terms within the brackets [ ] were reduced. The result, (36), is Einstein's famous photo-electric equation. It shows that the energy of an emitted photon varies linearly with its frequency f_a. The speed of interaction S_n is an independent natural constant. This speed has remained hidden, as a factor within Planck's constant, for over a century.[38] The radius λ_a is expressed in terms of the other quantities and it does not directly appear in the solution. The photo-electric effect (36) emerges with the action of these classical terms.

(36)

$$Energy = hf_a$$

Energy flows at points were the speed of sound (a longitudinal mechanical wave) equals the speed of light (a transverse electrical wave). These energy flows are observable. The point like photon appears as an observable effect of this energy flow. Light also acts like a disperse wave. These waves are observed indirectly with the appearance of many photons. The energy of a photon is a classical function of amplitude. This amplitude is expressed in volts. The frequency of the emitted photon f_a is that of the transitional quantum state. The frequency and the energy of the photon emerged in concert, from speed, within a subset of Newtonian mechanics.

Energy, Cold Fusion, and Antigravity

THE FINE STRUCTURE CONSTANT

Coulomb's equations (1&2) describe the electromagnetic force. The factors of Coulomb's equations were used by Maxwell to show that light travels at the luminal velocity **c**.

$$c = \sqrt{\frac{1}{e_0 u_0}}$$

Znidarsic's equations (3&4) also describe the electromagnetic force. The apparent speed of the standing wave S_n emerged from a fundamental analysis with equations (3&4). This speed was associated with Fermions. Equation (8) extended this relationship and showed that energy travels within quantum systems at a multiple of the speed $2S_n$. This speed was associated with Bosons. The ratio of the two traveling waves was presented below.

$$\alpha = \frac{2S_n}{c}$$

The origin and meaning of the fine structure constant is a long standing mystery. Richard Feynman stated,

"Physicists put this number up on their wall and worry about it."

The fine structure constant **α** materializes with the interplay of the transitional and the luminal speeds.

Energy, Cold Fusion, and Antigravity

THE DEBROGLIE WAVE

In 1924 Louis deBroglie published his thesis, "Recherches sur la théorie des quanta" (Research on the quantum theory). DeBroglie suggested that matter has wave like properties. The experiments of Davisson and Germer subsequently confirmed this hypothesis.[42] The length of the deBroglie wave varies inversely with its mass **M** and velocity *v*.

$$\lambda_d = \frac{h}{Mv}$$

The deBroglie wave is a curious construct. It shrinks and swells with speed, it has no classical analog, and it is based upon Planck's empirical constant. It is interpreted to be only a mathematical abstraction of probability. Schrödinger incorporated deBroglie's wave into his wave equation. It was said that Schrödinger's equation is what it is and that it cannot be reduced to fundamental terms. Despite these difficulties, Schrödinger's wave was accepted because it produced results. It describes most of physics and all of chemistry.

Louis deBroglie suggested that his wave is a real. He said that a beat note emerged from a superposition of the electron's Compton wave and its Doppler shifted reflection. The wavelength of the beat note is that of matter's deBrogle wave. The electron's Compton wave could not be produced from any fundamental analysis. There was no surface upon which to reflect the Compton wave. The result required

Energy, Cold Fusion, and Antigravity

matter to propagate at light speed. DeBroglie's beat note hypothesis was not well founded and it was promptly rejected.

This author has shown that an elastic discontinuity of capacitance C_q is fundamental to the Compton frequency f_c of matter. An extension of this analysis below presents the factor **n**, not as a description of a fine atomic orbit, but rather as a description of the downshifted Compton frequency of a condensate.

$$f_c = \frac{1}{2\pi} \sqrt{\frac{F_m / n^2 r_h}{M_{e^-}}}$$

The magnetic field can be attached to a magnet or it may fly away at light speed as a photon. What binds the force fields into the structure of matter? A professor explained that more advanced courses were required to address this question. Undeterred, this author noted that magnetic fields were pinned into type-2 superconductors at defects (discontinuities). The resulting lockup of the magnetic field is the reason why magnets stay afloat above these superconductors. The discontinuity r_p emerged from the spring like refactoring of Coulomb's equation. The discontinuity r_p pins the force fields into the structure of matter and it provides a reflective surface. The stability of matter is secured by a minimum of stray capacitance.

The phase speed of the Compton vibration is luminal **c**. The propagation of this luminal mode is restrained by an elastic discontinuity. An elastic discontinuity drags the Compton

Energy, Cold Fusion, and Antigravity

vibrations along at the group speed **v**. A good analogy, for the motion, is that of an imaginary very stiff bell. The electron's Compton wave propagates in the metal, of the bell, at the luminal speed. The electron's deBroglie wave emerges, as a beat note, with the bell's swing. Schrödinger's equation is not fundamental. It depends upon this classical action. The deBroglie wave was fully developed, as the consequence of a beat note, in the paper "The Constants of the Motion". This paper was published in "The Journal of New Energy" in the fall of 2000[36].

The concept of the elastic discontinuity of radius r_p is new and somewhat untenable. Perhaps it may be more palatable to view the discontinuity in terms of an effect of minimum of stray capacitance (1.5677×10^{-25} Farads). The gravitational force establishes this value of capacitance by its interaction with the extent of the universe. All points within the universe encounter this minimum of capacitive. The effect is similar to that of the isotropic capacitance of conducting sphere where all points within the sphere encounter the capacitance of the sphere. The electrical force experiences this capacitance through a maximum of elasticity. The strong nuclear force experiences this capacitance through a minimum of elasticity. The weak nuclear force resonates with this capacitance. The unitary nature of a minimum of stray capacitive establishes the quantum condition, determines the outcome of a quantum measurement, and it provides an absolute rotational frame of reference for mechanical motion.

Energy, Cold Fusion, and Antigravity

RELATIVITY

Einstein's theory of Relativity applies to the continuous properties of matter. Quantum theory applies to the individual particles of matter. There has been a long standing quest to establish these two theories from a single organizing principle. Erwin Schrödinger's time dependent equation is fundamental the contemporary depiction of the quantum realm. Schrödinger built the imaginary number *i* into the momentum operator (the left side) of his equation. The number *i* rolled matter's kinetic (magnetic) energy 90 degrees away from its potential (static) energy. This complex operation implicitly assumes that the real and imaginably components of the matter wave are orthogonal ⊥ to each other. Matter's energy oscillates between the real and imaginary domains within Schrödinger's wave equation. The total energy remains constant. The oscillations resemble those within the process of induction, induction acts orthogonally, and everything appeared to be perfect. Complex mathematics and Euler's equation were used to quantify the orthogonal relationship.

"Euler's equation is our jewel." Richard Feynman

The orthogonal structure was extended into an infinite number of dimensions within Hilbert space. Relativity did not naturally emerge from this model. The solutions had to be injected into the Relativistic Dirac equation in order to produce Relativistic results. Schrödinger's wave equation and Special Relativity were cobbled together without an organizing principle.

Energy, Cold Fusion, and Antigravity

A real standing wave emerged with this with this author's elastic constant. A real traveling wave emerged along with this author's wave number. These waves may be shown on a complex plain; however, both solutions are real and they reside within a Euclidean geometry. The superposition of the kinetic and the potential energies of these waves can exist at any phase ϕ. Relativistic observables emerge with the non-orthogonal alignment of the quantum superposition. No complex injections, imaginary abstractions, or special spaces are required.

Equation (9) describes the motion of the rest energy of a wavefunction. The motion is similar to that of a mass bouncing on a spring. When a spring is drawn tight the potential energy (i.e. electrostatic potential) is at a maximum and the kinetic energy (i.e. magnetic energy) is zero. One quarter of a cycle later the spring is relaxed and the kinetic energy is at a maximum. These waves have a high quantity factor and their amplitude is large. Matter's rest energy (**Mc^2**) is expressed by the length of the vertical vector on the phase plane (Pg. 129). The phase ϕ between the potential and the kinetic energy is expressed by the angular displacement of the vector.

Equation (5) describes the speed of the orbiting electron. This motion is that of a traveling wave. It is similar that of a surface water wave. The kinetic and potential energies flow in lock step along the surface. These waves have a low quantity factor and a low amplitude. The traveling component of matter's mass energy was placed along the **x** axis of a phase plain. Matter's relativistic momentum **P** is carried by the flow of its mass energy (**$P=E/c$**).

Energy, Cold Fusion, and Antigravity

Contemporary constructs describe two sources of momentum; that of moving matter **Mv** and that of flowing energy **E/c**. This author's construct has reduced to sources of momentum to one; the flow of energy. Observable reality emerges with this momentum along the **x** axis of the phase plane. For example, matter's kinetic energy equals the integral of this momentum with respect to velocity and the applied force equals the derivative of this momentum with respect to time. No quantum operators are required.

The vector sum $(x^2+y^2=z^2)$ of the standing and traveling energies equals the length of the hypotenuse **Z**. The hypotenuse represents matter's relativistic energy. The ratio of standing energy (along the **y** axis) to the relativistic energy (along the hypotenuse) reduces to:

$$(1-v^2/c^2)^{-1/2}$$

The ratio is known as the gamma function (γ). The arc sin of the ratio γ is the phase ϕ. This phase varies inversely with speed. Matter contracts as the phase between its potential and kinetic components decreases. This action appears as the Lorentz contraction of matter.

The "impedance matched" interpretation of quantum physics describes the real, classical motions of the matter wave. The elastic factor describes the motion that oscillates with respect to time. This motion was shown along the **y** axis of a phase plain. The wave number describes the motion that undulates with respect to position. This motion was shown along the **x** axis of a phase plain. Time dilates as the positional motion gains influence.

Energy, Cold Fusion, and Antigravity

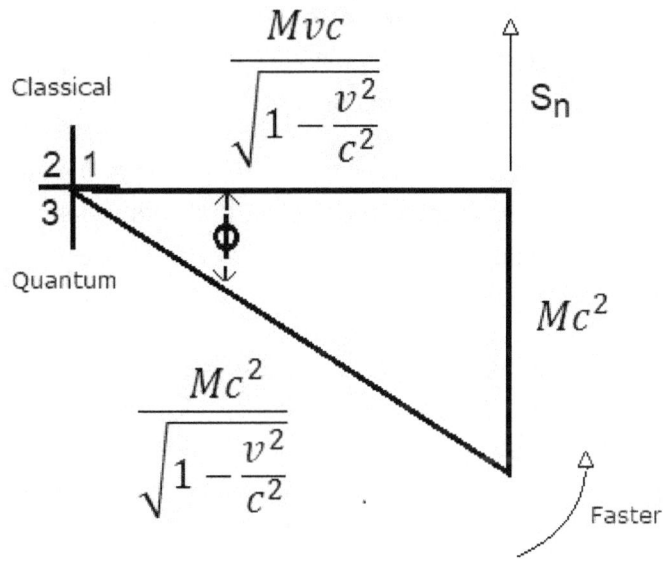

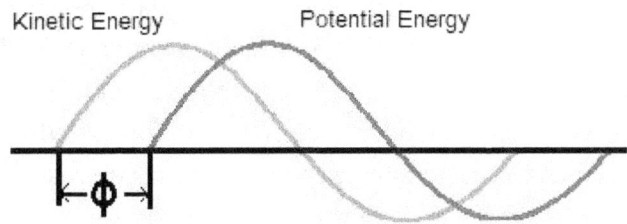

The angle between the potential
and the kinetic energy

The Standing and Traveling Waves of Matter

Energy, Cold Fusion, and Antigravity

Resting matter sets at a phase of 270 degrees. The longitudinal motions oscillate, like a mass on a spring, with matter under this condition. Znidarsic's analysis of (4) describes the rotational component of this velocity S_n. This is the velocity of a quantum collapse.

Energy flows at the luminal velocity at a phase of 0 degrees. Under this condition the induced electromagnetic field undulates with the flowing energy. Maxwell's' analysis of (2) describes the velocity **c** of this transverse motion.

The phase of matter, moving at sub-luminal speeds, is greater than 270 degrees and less than 0 degrees. The magnetic fields follow moving mass energy. Special Relativity emerges in combination with (2) and (4).

A great debate took place between Bohr and Einstein. Entangled quantum systems appear to react instantaneously. Special Relativity shows that information cannot travel faster than the speed of light. Einstein suggested that the entangled systems must contain hidden information. Entangled systems only seemed to communicate at superluminal speeds as this information expressed its influence. Bohr objected and stated that entangled systems do react instantaneously. This state of affairs remained unresolved until 1968 when John Bell proved that Bohr was correct and that quantum systems do react instantaneously.[38] How do quantum systems instantaneously react within the bounds of Special Relativity?[43] This author placed the bounds of Special Relativity in the third and the fourth quadrants of the phase plane. He suggests that these quadrants illustrate quantum behavior.

Energy, Cold Fusion, and Antigravity

At a conference in 1925 Peter Debye asked Erwin Schrödinger, "How can you have a wave without a wave equation?" Schrodinger then went on and developed his famous quantum wave equation. There are now two wave equations, one for the quantum realm and another for the classical realm. Schrödinger wave equation relates energy and the rate of change of this energy (frequency) with a first order differential equation. First order equations shift the phase of the motions (a sin becomes a cosine etc.). A complex factor of *i* was injected into the total energy (the right side) of Schrödinger's equation to roll the phase of the motions back into alignment. It was assumed, with this abstraction, that the quantum domain must be inherently complex.

"What is unpleasant and indeed directly to be objected to is the use of complex numbers. The wavefunction is surely a real function."   Erwin Schrödinger

First order equations produce singular solutions. Quantum uncertainty was introduced as a complex negative frequency.

This author developed a quantum solution that is based upon the classical wave equation. Velocities were related with an elastic constant and a wave number. The number of wave equations that are required to describe physical reality was reduced to one. The classical wave equation is entirely a second order differential equation. Second order equations (with no first order term) start with and maintain (sin to cosine and back again to –sin etc.) their phase alignment in space and time. Complex numbers are not fundamental to their solution.

Energy, Cold Fusion, and Antigravity

The classical wave equation produces pairs of solutions whose components are delimitated by a minus sign. This duality establishes the direction the velocity S_n which in turn quantifies Heisenberg's amount of uncertainty. The bidirectional spin of a quantum system emerges intrinsically with this solution. The left and right halves of the phase plane illustrate this duality. Complex mathematics can be used to efficiently describe the rotational motion but the action is inherently real. The juncture of Special Relativity, quantum physics, and observable reality is enumerated upon the phase plane.

The quantum condition emerges with a minimum of stay capacitance. Quantum systems react with this capacitance and flow with a negative phase (below the x axis on the phase plane). Electrical engineers refer to these negative phases as a leading power factor. The magnetic field, in an electrical line with leading power factor, travels ahead of the electrical field. The magnetic fields of moving matter cannot travel ahead of their static fields. This condition would put the cause before the effect and it would violate causality.

A second reversal of position is required. Superluminal speeds provide the second reversal. Causality is sustained within the quantum domain by the immediate action of its negative inertial mass. Spatial position has no meaning within the quantum domain; however, angular momentum is independent of position and it is universally conserved. No angular momentum can be exchanged with the quantum realm. This inactivity blocks the transfer of information which preserves causality, maintains Special Relativity, and inhibits genesis.

Energy, Cold Fusion, and Antigravity

The superimposed solutions of the classical wave equation carry no surplus of angular momentum. The ensemble of can interact within the quantum realm. Duality emerges as a second component of a paired solution remains within the quantum realm. Entanglement emerges with the simultaneous appearance of both solutions. This action is "spooky ". This process commences within the quantum domain and the outcome is unpredictable. The function (20) quantifies the magnitude, direction, and the duality of each transitional state. The exposition of these solutions is extraordinary and their unique existence suggests that mathematics was discovered.

The structures within the quantum domain immediately collapse upon the **x** axis during transition. Observable position and momentum appear along the **x** axis at the boundary of classical (upper) and quantum (lower) quadrants.

Energy, Cold Fusion, and Antigravity

This author has displayed matter's motions with polar coordinates on a phase plane. This geometry is Euclidian. The circular atomic orbits can be presented upon a Euclidian geometry. Physicists do not do espouse a Euclidian geometry for this purpose. They would rather display the atomic motions on a complex plain. They can then apply the complex product within a solution of the Schrödinger's wave equation. The complex product adds the angles and multiplies the amplitudes of two interacting vectors. A postulate is that a wave function's real and imaginary components are locked into an orthogonal orientation. These abstract, imaginary motions can display circular motions on a complex plane. The multiplication of the wavefunction with its complex conjugate within Hilbert space magically eliminates the imaginary components of the wave function as it extracts the square of a one dimensional displacement from an array of vectors. Max Born proposed that this displacement expresses the *probability of finding a particle*. Quantum discreetness was rudely imposed with the injection of Planck's constant. This imposition, like a nail driven into a piece of glass, shatters the solution into numerous dimensions. The destruction of the inertial mass of a collapsing wavefunction is not addressed. There is no physical mechanism to adjust the phase of the collapsing wavefunction. It is hard to believe that elementary particles can perform mathematical multiplication. The multiplication of two variables is a non-linear process and complicated solutions must be segmented within a fractured multi-dimensional vector space. The result is precise but the interpretation is abstract, fractured, non-relativistic, and incomplete.

Energy, Cold Fusion, and Antigravity

Electrical engineers perform a similar mathematical operation. They interpret the above the axis motions to be inductive and the below the axis motions to be capacitive. They superimpose the inductive and capacitive reactances with addition (not multiplication). Elementary particles can perform mathematical addition through a process of superposition. Addition is a linear process and complicated solutions emerge within Euclidian space. The relativistic energy is carried by the transitional state and resonates with the transitional velocity S_n. This energy is shown at 90 degrees up on the phase plane. This vibration is effected by the experimental boundaries. This influence appears as the observer effect. The relativistic energy collapses upon the x axis with the counterposition of the classical (above the x axis) and quantum vibrations (below the x axis). A minimum of incipient vibration activates a cascade. Excessive vibration quickens the reaction. Znidarsic proposes that the *probability of transition* is proportional to the square of the amplitude of the transitional vibration and varies inversely with the strength of the bond. Particles decohere from their bonds with this vibration. Quantum discreteness emerges with the harmonics of the motion. This vibration reduces the quality factor of the matter wave, lowers its amplitude, and destroys its inertial mass. The impedance of the system is matched and power flows without bounce. Real power flows along the **x** axis at zero degrees with the collapse of the quantum wave function. Capacitance **Cq** opens a path for the phase adjusting action. The wavelike nature of these observables introduces an uncertainty in position. This interpretation is precise, real, linear, relativistic, and complete. The action bares the collapse of a quantum wave.

Energy, Cold Fusion, and Antigravity

General Relativity applies universally and it has been difficult to identify a single causative action. Newton described the action in terms of a force field. Einstein described the action it in terms of a time dilatation. Time is metered by matter's Compton frequency. The lower frequency of the LENR emission spectra can be interpreted to be an effect of the downshifting of the Compton frequencies of the nucleus. The strong, weak, and electromagnetic forces are adjoined into the resulting temporal distortion. The increased LENR reaction rate is simply due to the intense stimulation that is coming in from normal time. This slowing of local time appears independently outside of the bounds of General Relativity. This independent action, when applied globally, provides an insight to the juncture of gravity and matter at the quantum scale. A minimum of stray capacitance is established as the gravitational field weakly interacts with the extent of the universe. The other forces experience this capacitance through contact with matter. This minimum of this stray capacitance increases in the proximity of other mass. Matter softens, the Compton frequency of matter downshifts, and time dilates. These actions propel the elastic bubble r_p in a direction that increases the intensity of the composite gravitational field. Gravity binds separate quantum structures into the continuity of the quantum realm through the imposition of a unitary minimum of stray capacitance.

The escape velocities of the Milky Way galaxy is approximate the transitional velocity S_n. This is probably only a coincidence.

Energy, Cold Fusion, and Antigravity

A charge's strength and its spin are independent of the structure or type of a particle. This commonality suggests that the source of charge and spin does not originate within mater. This author suggests that the electric charge originates as the capacitive discontinuity r_p disrupts universal, uncharged background field. This concept is similar to John Wheeler's single electron postulate; however, it requires no electrons traveling backwards in time.[47] The static fields appear as a point like ruptures while the magnetic components have structure at the larger radius r_p. The expulsion of magnetic flux from a particle to the elastic radius standardizes the states of quantum spins, it supports quantum oscillations, and it sets the Fermi displacement length of the nucleus. There is way to test and determine which model is correct. This author's model suggests that an asymmetrical discontinuity would generate a transitional force. Indeed, Dr. Charles Buhler of Exodus Propulsion Technologies claims to have detected such a force. The state of charge and spin emerge from the imposition of an elastic discontinuity. The action provides an absolute rotational frame of reference for fields and it is fundamental to the conservation of charge.

The cosine of the phase ϕ expresses the power factor of an electrical power line. The power factor is the ratio of the flowing energy (measured in watts) to the standing energy (measured in vars). The magnetic reactance of a power system is not conservative. Freewheeling synchronous motors are sometimes attached to a power line in order to cancel the magnetic reactance of a power line. The power factor of the line then varies with the phase of its motion.

Energy, Cold Fusion, and Antigravity

Correspondingly, the inertial mass of matter is not conservative. A strong gravitomagnetic field that is induced by a rotating, vibrating superconductive disk of the opposite handedness may cancel the inertial mass of matter. The reduction may make other propulsion systems more productive. The momentum of this matter will at first be conserved by moving faster. It may be possible to continue this process until the inertial mass of the matter becomes zero or negative. Can a mass with no inertial mass travel faster than the speed of light without going backwards in time? Can a spacecraft to travel at superluminal speeds? Do U.A.P.'s fly this way?

This author's refactored Coulomb's equation (3) describes a force. This force has both a magnitude and a direction. Force is a Newtonian quantity. Equation (4) was produced as a result of the integration of Equation (3). Equation (3) describes an energy. This energy has a magnitude but it does not have a direction. Energy is a scalar quantity. A reason that this author used Equation (3) is because scalar computations are simpler. The direction of a force does not have to be vectored into the solution. Quantum physicists simplify their quantifications in a similar fashion. They use Hamiltonian operators. These operators have no force or direction. They are scalar. Schrödinger's wave equation is scalar. There is no mechanism to retrain a real wave. The properties of Relativity do not spontaneously emerge without restraint. This author's Equation (3) shows that mass energy is restrained at elastic discontinuities. The properties of Relativity, matter's deBroglie wave, the momentum of matter, the superpositions of states, and the condition of the

Energy, Cold Fusion, and Antigravity

quantum realm have emerged from an analysis of the group velocity of a restrained wave.[40]

The standard model describes most of physics. It emerged as an effect of Planck's constant. The static components of the force fields dominate and the magnetic components are dispatched. The model was extended to the energy of excited quarks on Feynman diagrams. The diagrams depict an exchange of virtual particles. This model becomes fuzzy at classical energies. Most importantly, this model describes the quantum reaction in terms of its products.

Znidarsic's interpretation is not based upon Planck's constant. The transitional quantum state, not virtual particles, mediates the reaction. The magnetic components of the force fields are of a determining consequence. Most importantly, this model describes the quantum reaction in terms of its path. This new interpretation would confound the scientific establishment and there would have to be good reasons to make use of it. Measurement is a transitional process and the description of its path clarifies the measurement problem. The model extends upward only to the level of relaxed quarks and may reveal noting more than the standard model at higher energies. The interpretation does; however, provide a unique vantage point. The perspective extends clearly downward into the classical realm. The nature of low energy exchange reactions was exposed. The technologies that emerge from this exposition are imperative.

This author appreciates that his model is nascent; however, it does clarify many hidden aspects of reality.

Energy, Cold Fusion, and Antigravity

CHAPTER 8 Generations

J.J. Thomson discovered the electron 1897. A heavier version of the electron, the muon, was discovered by Carl D. Anderson and Seth Neddermeyer in 1936. This discovery was quite unexpected and Nobel laureate Isidor Isaac Rabi famously quipped, "Who ordered that?" A still heaver version the electron the tau was discovered in the mid-1970s by a group led by Martin Perl. These three forms of the electron are known as generations. A similar set of generations exists for quarks. The standard model of physics examines resonances within the stationary quantum states. This model has offered no explanation as to why the three generations of particles exist.

A clue appears with the stability of the ground state of the atom. The electron cannot cohere below the ground state because the boundary conditions below this level suppress the transitional state. A second clue appears with the double slit experiment. Particles decohere at the slit because the boundary conditions at the slit enhance the transitional vibrations. The electron passes through the slits as a wave in a state of quantum transition. A third clue appears with the stability of the bound neutron. An accumulation of transitional vibration increases the isospin of the nucleus and activates the reaction. This author has searched for boundary condition that provides for muon stability. This understanding may expedite the development of materials that inhibit the decoherence of the muon (double slit in reverse) and lead to the commercialization of muon activated fusion.

Energy, Cold Fusion, and Antigravity

CHAPTER 9 Conclusion

Coulomb's equations quantified the strength and the energy of an electrical field. Maxwell extended Coulomb's equations and determined the speed of light. Planck's constant *h* was the oddball. It did not emerge from Maxwell's formulations. It is empirical. Planck's constant was injected, abstractly, into the contemporary description of the quantum domain. It was assumed that Newtonian mechanics exists within a subset of this reality.

Znidarsic refactored Coulomb's equation into the form of a spring. His construct also quantified the strength and the energy of an electrical field. A wave-like elastic factor K_e- and a particle-like elastic discontinuity r_p emerged as terms within the refactored equation. The discontinuity was associated with a minimum of stray capacitance. The stationary quantum states are affixed at this discontinuity. The length $2r_p$ was used to set the wavelength λ of the electrical field's wave number ($2\pi/\lambda$). The speed of sound in a solid and the spins of the electron naturally resonate with this elastic factor and wave number.

The process of low energy nuclear reactions and the process of strong gravitomagnetic induction have been held in distain. A speed was observed during these experiments. This speed corresponds with the speed of sound in the nucleus. The analysis of this speed gave the atomic energy levels and produced the photoelectric effect. The duality of particles and waves appeared concurrently. The quantum realm emerged within a subset of classical mechanics.

Energy, Cold Fusion, and Antigravity

The quantum, classical, and relativistic conditions were derived as functions of the classical wave equation. Quantum uncertainty appeared with the pairing of these functions. This result is real, not complex. The model's motion constants apply seamlessly at all scales and no abstractions were invoked. Planck's regional constant and Schrödinger's wave equation are not required. The non-conservation of inertial mass is central to the arguments. It was shown that the inertial mass of matter is destroyed with the compression of a surface wave. The continuous, harmonious, and symmetrical formulation is beautiful.

Taken individually each result can be contested. Taken together the model's range extends smoothly through fifteen orders of magnitude in spatial dimension. This scope implies that the model can be applied, within this domain, to make useful predictions. An interpretation of the result shows how to the control all of the natural forces.

The process of vibration may adjoin the nucleons, within a superconductor or proton conductor, into a new state of matter. This state is one of a continuous quantum transition. Trillions of atoms will be adjoined into a single state of transition. Strong, local gravitomagnetic effects will emerge. The use of these strong gravitomagnetic forces will propel space craft to the stars. Long range nuclear magnetic effects will emerge. These nuclear-magnetic forces can be used to reduce nuclear waste and to supply an unlimited amount of clean nuclear energy. Man may be on the verge of harnessing all of the natural forces. This technology will change the type of our civilization. I hope that this happens soon.

Energy, Cold Fusion, and Antigravity

ABOUT THE AUTHOR

Frank Znidarsic is a registered professional Electrical Engineer in the state of Pennsylvania. He graduated, in 1975, with a B.S. in Electrical Engineering from the University of Pittsburgh at Johnstown. In 1987, he went on to obtain an A.S. in business administration from Saint Frances College. He studied physics at the Indiana University of Pennsylvania in the late 1990s. He has worked in industry for 32 years. He has personally witnessed the decline of American manufacturing as his position has been downsized or outsourced at least four times. He has been told more than once, "*As one door closes another opens.*"

Frank has written peer reviewed and non-peer reviewed articles on low energy nuclear reactions, antigravity, and physics. These works have shown a way to produce limitless quantities of energy. They have also shown how to propel spacecraft to the stars. The emergence of these technologies, along with the continuous development of computer science, will lead to a future age of abundance. Frank believes that low level jobs will be lost to automation and that new, exciting opportunities will emerge with our cosmic expansion. Frank hopes that the introduction of this book will expedite the next technological revolution.

Address comments to:

fznidarsic@aol.com

Energy, Cold Fusion, and Antigravity

NOMENCLATURE

BTU =3412 kilowatt-hour

c = 299,792,458 meters/second. Light speed

f_a = The atomic frequency

f_c = 1.236 x 10^{20} Hertz, The Compton frequency of the electron

F_m = 29.0535 Newtons, The limit of the electron's force

G = 6.67 x 10^{-11} The gravitational constant

h = 6.625 x 10^{-34} joules-seconds Planck's constant

K_{e^-} = F_m/r Newtons/meter, The elastic factor of the electron

K_f = 1.35969 x 10^{-15} meters, The nuclear spacing

M_{e^-} = 9.109 x 10^{-31} kg, The mass of the electron

M_p = 1.6726 x 10^{-27} kg, The mass of a proton.

n = 1,2,3,4,5.., an integer

Q = 1.60218 x 10^{19} Coulombs, The electric charge

r_a = The atomic radius

r_p = 1.40897 x 10^{-15} meters; The radius of a particle.

$2r_p$ = 2.81794 x 10^{-15} meters, The classical radius of the electron

r_h = .529 x 10^{-10} meters, The radius of the hydrogen atom

S_n = 1,093,850 meters per sec., The speed of sound in the nucleus

S_a = The speed of an atomic electron

Z = 1,2,3,..., The number of protons in the nucleus

Energy, Cold Fusion, and Antigravity

THEOREMS

1. The constants of the motion tend toward the electromagnetic in a Bose condensate that is vibrated at a dimensional frequency of 1,093,850 meters per second.

2. The impedance matched interpretation of quantum physics states that **2πn** times the speed of a transitional electron equals a harmonic **Z** of the fundamental nuclear speed S_n

3. Bosons bounce. Their speed is expressed in meters/second.

4. Fermions fail. Their motion is expressed in megahertz-meters.

5. Quantum systems exist within a domain of negative inertial mass.

6. The quantum domain emerges, within a subset of Newtonian mechanics, as an effect of minimum of stray capacitance.

7. The compression of a surface wave destroys internal mass.

Energy, Cold Fusion, and Antigravity

PARTIAL BIBLOGRAPHY

1. I. Bernard Cohen, Henry Crew, Joseph von Fraunhofer, De Witt Bristol Brac, "The Wave theory, light and Spectra", Ayer Publishing, 1981
2. Robert Bunsen, article, Journal of the American Chemical Society, Volume 22, 900
3. L Hartmann, Johann Jakob Balmer, "Physikalische", Blätter 5 (1949), 11-4
4. W. Ritz, "Magnetische Atomfelder und Serienspektren", Annalen der Physik, Vierte Folge. B and 25, 1908, p. 660–696.
5. Planck Max, "On the Law of the Distribution of Energy in the Normal Spectrum", Annalon der Physik, Vol. 4, p 553, (1901).
6. Einstein Albert, "Development of our Conception of the Nature and Constitution of Radiation", Physikalische Zeitschrift 22, (1909)
7. Bohr Niels, "On the Constitution of Atoms and Molecules", Philosophical Magazine, Series 6, Vol. 26, pp 1-25 (1913)
8. Maxwell James Clerk, "A Dynamical Theory of the Electromagnetic Field", Philosophical Transactions of the Royal Society of London, Vol. 155, (1865)
9. Louis deBroglie, "Recherches sur la théorie des quanta" (Researches on the quantum theory), Thesis, Paris, 1924
10. Max Born, "The Statistical Interpretation of Quantum Mechanics", Nobel Lectures, 1964
11. A Einstein, B. Podolsky, and N. Roses, "Can Quantum-Mechanical Description of Physical Reality Be Considered Complete", Phys. Rev. 47, 777 - 780 (1935)
12. Miley George H., "Nuclear Transmutations in Thin-Film Nickel Coatings Undergoing Electrolysis", 2nd International Conference on Low Energy Nuclear Reactions, (1996).
13. Mosier-Boss, Szpak S., Gorden F.E. and Forsley L.P.G., "Use of CR-39 in Pd/D co-deposition Experiments", European Journal of Applied Physics, 40, 293-303, (2007)
14. Storms Edmond, "Cold Fusion, A Challenge to Modern Science", The Journal of Scientific Exploration, Vol 9, No. 4, pp 585-594, (1995)

15. Rothwell Jed, Infinite Energy, Issue 29, p 23. (1999) "50 nanometers ..is the magic domain that produces a detectable cold fusion reaction"
16. Arata Y. and Fujita H., Zhang Y., "Intense deuterium nuclear Fusion of Pycnodeuterium-Lumps Coagulated Locally within highly Deuterated Atomic Clusters", Proceedings of the Japan Academy, Vol. 78, Ser.B, No.7 (2002)
17. Li Ning and Torr D.G., "Gravitational effects on the Magnetic Attenuation of Superconductors", Physical Review B, Vol 46, #9, (1992)
18. Reiss Harrald, "Anomalies Observed During the Cool-Down of High Temperature Superconductors", Physics Essays, Vol. 16, No. 2 (June 2002).
19. Tajmar M., deMathos C, "Coupling of Gravitational and Electromagnetism in the Weak Field approximation", Cornell
20. Podkletnov E. and Levi A.D., "A Possibility of Gravitational Force Shielding by Bulk YBa2Cu3O7-x, Superconductor", Physica C, vol 203, pp 441-444 (1992).
21. Podkletnow and Giovanni Modanse http://www.scribd.com/doc/39624178/Podkletnov-2001-Paper-with-Giovanni-Modanese
22. Papaconstantopoulus D. A. and Klein B. M., "Superconductivity in Palladium-Hydrogen Systems", Phys. Rev. Letters (July 14, 1975)
23. M. Modarres, Momentum "Distributions of Nuclear Matter", 987 Europhys. Lett. 3 1083
24. A. Sommerfeld, "Principles of the Quantum Theory and the Bohr Atomic Model", Naturwissenschaften (1924), 12 1047-9
25. Richard Feynman, "The Strange Theory of Light and Matter", (1988)
26. The Lex Foundation, "What is Quantum Mechanics", (1996)
27. For over 2,000 papers on cold fusion; LENTR. ORG
28. Matthew R. Simmons; "Twilight in the Desert"
29. Frank Znidarsic "The Elastic Limit and the Quantum Condition", "The General Science Journal"
30. Todd Tucker, "Atomic America", 2009
31. Sean Carroll, "From Eternity to Here", Pgs 178, 242

32. Letters "Ernest Rutherford to Neils Bohr", "Yale University Press", page 68, (1938)
33. Charles P. Poole, "Superconductivity", Pg. 4 (2007)
34. MA Preston and Rajatk Bhadon, "The Structure of the Nucleus" Westview Press page 248-252 (1975)
35. Mark L. Stockman, "Nanoplasmonics, The Physics Behind the Applications", Physics Today, (2/ 2011)
36. Frank Znidarsic, "The Constants of the Motion", The Journal of New Energy, Vol 5. No 2, Sept 2000 & http://www.padrak.com/ine/NEN_5_11_12.html
37. I. Sick ; "Precise Radii of Light Nuclei from Electron Scattering," Springe-Verlag (2008) http://www.sjsu.edu/faculty/watkins/He4nuclide.htm
38. BELL, Department of Physics, University of Wisconsin, Madison, ON THE EINSTEIN PODOLSKY ROSEN PARADOX, Physics Vol. 1, No. 3, pp. 195-290, 1964
39. R C Jennison, Relativistic phase-locked cavities as particle models, Journal of Physics A: Mathematical and General, Volume 16, Number 15, 1983
40. F. Znidarsic, Force and Gravity, Infinite energy, Issue 22,October-November 1998
41. Many authors, Nonlinear lattice dynamics as a basis for enhanced superconductivity, Nature, Vol. 516, Issue 7529, 2014 http://www.nature.com/nature/journal/v516/n7529/full/nature13875.html#ref8
42. Davisson, C. J.; Germer, L. H. (April 1928)."Reflection of Electrons by a Crystal of Nickel", *Proceedings of the National Academy of Sciences of the United States of America.* (14)(4):
43. Antonio Gianfrate; Superluminal X-waves in a polariton quantum fluid, Nature, Light: Science & applications volume7, page 17119 (2018)
44. LeeAnn M. Sager, Shiva Safaei, and David A. Mazziotti Physical Review B Potential coexistence of exciton and fermion-pair condensations (2020)
45. E.E. Podkletnov, Moscow Chemical & Scientific Research Center Weak gravitation shielding properties of composite bulk $YBa_2Cu_3O_7$ Superconductor below 70K under e.m. field (1997)

Energy, Cold Fusion, and Antigravity

46. M R Staker, Department of Engineering, Loyola University Maryland, <u>A model and simulation of lattice vibrations in a superabundant vacancy phase of palladium–deuterium</u> (2020)
47. Johns Wheeler stated that all electrons are the same electron moving backward and forward in time in the 4th dimension.
48. P.L. Hagelstein / Journal of Condensed Matter Nuclear Science 35 (2022) 49–65

ADDITIONAL READING

Peer reviewed publication:

Frank Znidarsic, "<u>The Quantum Condition and an Elastic Limit</u>"; Bentham Open Chemistry Journal, ISSN: 1874-8422, Volume 1, 2014

Android Apps at Amazon and Google:

- The Dreams Alarm Clock
- The Heckler
- MIDI Staff
- Monitoring-MIDI,
- Parrot Teacher,
- Video Selfie Start/Pause/Stop
- Video Selfie Start/Stop
- Voice Staff

Scan the QR code above with your cell phone camera to view my products.

www.ingramcontent.com/pod-product-compliance
Lightning Source LLC
Chambersburg PA
CBHW061510180526
45171CB00001B/116